Suziane Alves Josino Lima
Antônio Vitor Machado

Quality of coconut water sold in the hinterland of Paraíba and Ceará

Suziane Alves Josino Lima
Antônio Vitor Machado

Quality of coconut water sold in the hinterland of Paraíba and Ceará

Physical-chemical, microbiological and sensory evaluation of green dwarf coconut water sold in the Paraíba and Ceará hinterlands

Imprint

Any brand names and product names mentioned in this book are subject to trademark, brand or patent protection and are trademarks or registered trademarks of their respective holders. The use of brand names, product names, common names, trade names, product descriptions etc. even without a particular marking in this work is in no way to be construed to mean that such names may be regarded as unrestricted in respect of trademark and brand protection legislation and could thus be used by anyone.

Cover image: www.ingimage.com

This book is a translation from the original published under ISBN 978-613-9-61714-2.

Publisher:
Sciencia Scripts
is a trademark of
Dodo Books Indian Ocean Ltd. and OmniScriptum S.R.L publishing group

120 High Road, East Finchley, London, N2 9ED, United Kingdom
Str. Armeneasca 28/1, office 1, Chisinau MD-2012, Republic of Moldova, Europe
Printed at: see last page
ISBN: 978-620-7-70022-6

Copyright © Suziane Alves Josino Lima, Antônio Vitor Machado
Copyright © 2024 Dodo Books Indian Ocean Ltd. and OmniScriptum S.R.L publishing group

SUMMARY

SUMMARY

Brazil is currently experiencing a growing trend in the cultivation of green dwarf coconut palms, which are distributed practically throughout the country. Coconut water is used in popular culture as a substitute for water and also to replenish electrolytes in cases of dehydration. The aim of this study was to evaluate the storage process of green dwarf coconut water produced commercially and bottled aseptically by industries in the Sertao Paraibano and Ceará. Samples were collected and identified from two industrial units on the day they were produced, transported to the laboratory in isothermal boxes with ice, where they were analyzed for physical, physical-chemical, microbiological and sensory parameters in order to assess the quality of coconut water immediately after processing. The samples obtained from the factories were stored according to experimental factorial design 22 with three repetitions of the central point, in order to quantitatively evaluate the influence of the input variables on the responses: turbidity, conductivity, viscosity, pH, total soluble solids, acidity and ascorbic acid. It was found that refrigerated coconut water should be stored at low temperatures so that its shelf life can be ten days. With storage time, there was a decrease in the pH, vitamin C, electrical conductivity and total soluble solids values for both industries, with the total soluble solids values being in line with what is permitted by legislation, except for experiment 4.The pH of all the experiments is not in line with what is allowed by legislation, there were simultaneous increases for the industries in the analyzed parameters of turbidity, viscosity, and total titratable acidity, and the time x temperature binomial considerably influenced the analyzed parameters during storage. Microbiologically, the factories do not comply with the standards established by current legislation for thermotolerant coliforms, molds and yeasts. No *Salmonella sp* values were found in any of the samples analyzed. In terms of taste, the coconut water was unfit for human consumption after 15 days in storage, with altered organoleptic characteristics. It is essential to implement and monitor Good Manufacturing Practices for industries that bottle coconut water.

Keywords: *Cocos nucifera L,* preservation, refrigeration.

1. INTRODUCTION

There is a wide variety of tropical fruits, but only a small number of them are grown and processed on a large scale. This is mainly due to the lack of infrastructure and the low level of technical knowledge (LIRA, 2010).

Brazil is the world's third largest fruit producer (FERRAZ, 2009) and is currently experiencing a growing trend in the cultivation of dwarf green coconut palms, which are distributed practically throughout the country. Although coconut cultivation is being stimulated and introduced in various regions of the country, the largest plantations and production are concentrated in the coastal strip of the Northeast and part of the Northern region of Brazil Embrapa, (2011). Favored by tropical climate conditions, both regions (North and Northeast) hold close to 70% of Brazil's coconut production. Brazil has around 280,000 hectares under coconut cultivation, spread over almost the entire national territory, with production equivalent to two billion fruits. Although there has been an increase in the area harvested since 1990, there has been a vertiginous increase (over 800%) in production since the end of the 1990s (FAO, 2011).

Ceará is the third largest coconut producer in the country, behind only Bahia and Parà Toda Fruta, (2011). The fruit of the coconut palm and its by-products are among the state's main export products (NORDESTE RURAL, 2009).

The state of Paraiba is the seventh largest producer of coconut (*Cocos nucifera* L.) in the Northeast. The area planted is concentrated mainly in the Litoral Norte, Litoral Sul and Joao Pessoa (dry coconut) and Sertao (green dwarf coconut) micro-regions (IBGE, 2007).

Since the 1990s, as people became more aware of the benefits of natural foods, there has been an increase in the exploitation of the dwarf coconut palm with a view to producing the green fruit for water consumption, which is a natural product with excellent nutritional qualities (EMBRAPA, 2010).

Coconut water is a liquid from the endosperm and corresponds to 25% of the weight of the fruit, where the amount of water can vary from 300 to 600 mL of coconut. It contains a variety of nutrients and is a good source of important minerals, such as magnesium, calcium and potassium, which, together with sugars, give it a pleasant taste, making it a natural isotonic drink (SEREJO et al., 2010).

Coconut water is used in popular culture as a substitute for water, as well as to replenish electrolytes in cases of dehydration Aragao et al., (2001).Traditionally, it is sold inside the fruit itself, a practice that involves problems related to transportation, storage and perishability of the product. Its industrialization is of fundamental importance, as it allows it to be consumed in places outside the producing regions, with the aim of reducing the volume and weight transported and, consequently, costs, as well as increasing its shelf life (ABREU, 2008).

A more recent and rapidly expanding type of processing is the extraction and packaging of coconut water (liquid endosperm) through the application of processing and preservation technologies. There are basically two methods of preserving bottled and refrigerated green coconut water (ROSA; ABREU, 2000).

Refrigerated coconut water is sold in "PET" (polyethylene terephthalate) plastic packaging, but you can also use cups with heat-sealable lids or low-density polyethylene (LDPE) bottles. The filling stage should be carried out in the shortest time possible, preferably with the product pre-cooled. The storage temperature should be kept at around 5 to 8° C, according to (ROSA; ABREU, 2000).

In general, small industries do not have qualified people and specific knowledge about the methods used to improve food safety, thus facing public health problems (FAO; WHO, 2007).

Consumers expect risk protection to be part of the entire production chain, from agricultural production to the consumer. Protection will only take place if all sectors of the production chain and inspection bodies act in an integrated manner, and if food control systems take into account all stages of the production chain (FAO; WHO, 2007).

Knowing that coconut water is a natural solution rich in mineral salts, sugars, vitamins and proteins, we suggest a study of the quality of green dwarf coconut water after processing, with the aim of providing industries that bottle coconut water with better refrigerated storage without any loss of product. Evaluating the physico-chemical, microbiological and sensory quality of ana verde coconut water, commercially produced by industries in the sertao of Paraiba and Cearà at three refrigeration times and temperatures, identifying the best temperature and storage time so that there are no losses in quality, with an exact definition of how coconut water bottling industries should operate, within a very promising contingent.

1.1 - General objective

To evaluate the physico-chemical, microbiological and sensory quality of green dwarf coconut water produced commercially by industries in the sertao of Paraiba and Cearà.

1.1.1 - Specific objectives

- Determine the physico-chemical and microbiological characteristics of commercialized anas verde coconut water, assessing its nutritional quality;

- Evaluate the quality of anao verde coconut water in terms of sensory parameters;

- To verify the influence of the time x temperature binomial in the refrigerated storage condition through a factorial experimental design on the quality of processed and commercialized coconut water;

•	Based on the results obtained, possible flaws in processing, transportation and storage were identified, with a view to improving the quality of green dwarf coconut water produced commercially by industries in the Sertao region of Paraiba and Cearà.

2. literature REVIEW

2.1 Coconut tree

2.1.1 History

According to Loiola (2009), due to the lack of direct evidence, *there is* no *way of* reporting the center of origin of the coconut palm, mainly due to the non-existence of its common ancestor. Based on circumstantial evidence, the most accepted hypothesis among researchers is that the coconut tree originated on the islands of Southeast Asia, between the Indian and Pacific Oceans. From there, it was transported to East and West Africa, from where it was introduced to the Americas.

The author also reports that the coconut palm was introduced to Brazil in 1553 in Bahia, with the Gigante variety coming from the island of Cape Verde. The Anao coconut tree was introduced at the beginning of the 20th century, with the Anao Verde cultivar coming from Java in 1925; the Anao Amarelo in 1938, from northern Malaysia; and the Anao Vermelho, from northern Malaysia, in 1939.

The coconut palm is considered to be the tropical species of greatest socio-economic importance in the intertropical regions, due to the plant's versatility of use. It plays a major social role, especially in coastal regions, where it is mostly cultivated by small producers on poor, sandy soils that are unsuitable for other types of activity Chan; Elevitch, (2006). The countries that stand out in terms of commercial coconut cultivation are: Indonesia, the Philippines, India, Brazil, Sri Lanka, Thailand, Mexico, Vietnam, Papua New Guinea and Malaysia (FAO, 2009).

2.1.2 Botanical aspects

The coconut palm belongs to the following botanical systematics: Kingdom: Vegetable; Branch: Phanerogams; Sub-branch: *Angiosperms;* Class: *Monocots*; Order: *Principes*; Family: *Palmàce* a; Genus: *Cocus* ; Species: *Nucifera*. With the scientific name of Cocus *nucifera* L. (IDESP, 1975).

The name *Cocos nucifera*, according to Benassi (2006), refers to the plant that produces nuts with the appearance of a head, as the word "coco" comes from the Portuguese "cabeça" meaning head and "nucifera" from the Latin "nucifer-a-um", which produces nuts.

The *Cocus* genus consists solely of the species *Cocus nucifera* L. This species is made up of a number of varieties, the most important of which, from an agronomic, socio-economic and agro-industrial point of view, are the Typica (Giant variety) and Nana (Ana variety) varieties. The Gigante variety is widely exploited, mainly by small producers, and currently accounts for around 70% of coconut plantations in Brazil. The Ana variety is made up of the Yellow, Green and Red cultivars, which are of little commercial importance (BHATTACHARYYA; BHATTACHARYYA, 2002).

The coconut palm is a perennial plant, a smooth-stemmed palm that can reach up to 25 m in height and 30 to 50 cm in diameter, with this height varying according to the age and variety of the plant. The leaves originate from a single growth point and are arranged in a spiral. They are wide and long with an average annual production of around 21 leaves, which are up to 4 m long and have 200 to 300 leaflets (GOMES, 2003).

It has an inflorescence that grows in the axils of the lower leaves, protected by bracts, also called swords, which are up to 70 centimeters long and develop in three to four months. The number of female flowers on each inflorescence is an indicator of productivity, varying according to genotype and environmental conditions, but it cannot be said that plants with a higher number of female flowers must necessarily have the highest yield. An inflorescence spends two-thirds of its life inside the plant and only one-third outside it, so productivity is influenced by the prevailing soil and climate conditions at the time of flower initiation (SANTOS FILHA, 2006).

The fruit of the coconut palm, the coconut, as can be seen in Figure 2.1, is formed from the outside in, by a smooth epidermis or epicarp, which is a thin film that surrounds the outside of the fruit and encloses the mesocarp, which is thick, fibrous and brownish, when dry, and is of great industrial use. The fruit is known as a large nut with a seed covered in a hard shell. Inside the shell is the kernel, which is the edible part, about 1 cm thick, with a cavity filled with liquid - coconut water (FERREIRA et al., 1998), which is slightly acidic and very rich in phosphorus and potassium. This liquid consists of approximately 93% water (MEDINA et al., 1980).

Figure 2.1 - Coconut palm fruit.

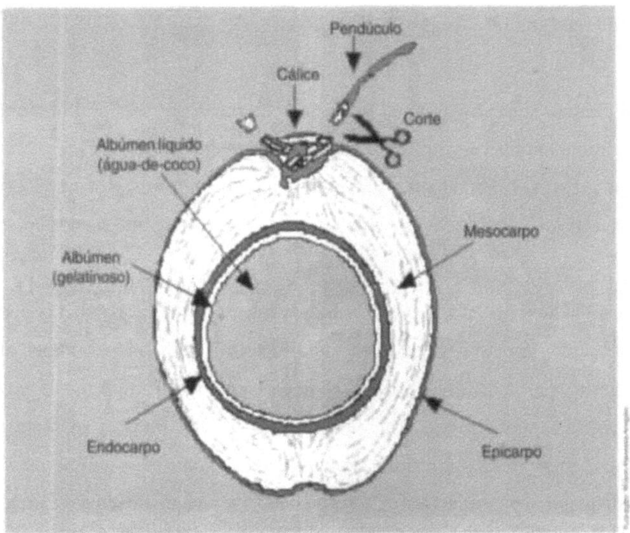

Source: Aragao et al., 1998.

The fruit is fully ripe from the twelfth to the thirteenth month. At this stage it is made up of around 35% mesocarp, 12% peel, 28% pulp and 25% water (SREBERNICH, 1998).

Coconuts reach their maximum size and weight at around six months of age, regardless of the variety. Both remain constant for one or two months. After this period, the weight decreases a lot and the size decreases slightly. In the last stage of ripening, the water that used to completely fill the inside of the kernel decreases due to evaporation or absorption by the solid albumen, which is what contributes to the greater weight loss. However, in very young coconuts, the solid albumen is completely missing and then appears, tender, thin and delicate, lining the inside of the endocarp wall (ARAGÂO et al., 2002).

The solid core begins to form around the fifth month. At this stage, the pulp has a gelatinous consistency, becoming rigid and reaching its maximum weight between 11 and 12 months, when it is used for culinary and agro-industrial purposes (ARAGÂO et al., 2002).

The pulp of the Anao coconut cultivar is preferred as a raw material for the agro-industrial manufacture of "light" coconut-based foods, and for culinary use in the preparation of low-fat foods, as the Giant coconut has higher fat content than the Anao (LOIOLA, 2009).

The coconut palm is a tropical palm, and its development is favorable in warm and humid climates, which are found between latitudes 20° N and 20° S. A temperature of 27 °C is considered optimal for the dwarf coconut palm, which has its development impaired if the daily minimum temperatures are lower than 15 °C (BRASIL, 2000).

2.1.3 Ana Verde Variety

The ana variety probably originated from a gene mutation in the Gigante variety. It is the coconut palm variety most commonly used commercially in Brazil for coconut water, although it can also be used in the agro-industry for food and/or dried fruit in natura. Due to its financial profitability and the growing consumption of coconut water in large urban centers, producers have become interested in the crop (MACIEL, 2008).

The Anao coconut tree comes into production at 2.5 years of age, producing an average of 200 fruit plants per year. Although yearling coconuts produce earlier than giant coconuts and have good yields, the pulp tends to be softer and more flexible, and of lower quality than giant coconuts, which are used exclusively for coconut water. The Anao coconut tree is slow-growing and precocious, flowering two years after planting. It can reach a height of 10 to 12 m and has a lifespan of around 30 to 40 years. It has a slender stem, numerous but short leaves and produces a large number of small fruits (150 to 200 fruits/plant/year), provided that the correct technology is applied when growing it. It is more sensitive to attacks from pests and leaf diseases, but it is the most tolerant of unfavorable environmental conditions, and the most similar to the Giant coconut palm (CHAN; ELEVITCH, 2006).

The period between fruit formation and ripening is around twelve months. The flowering season in Brazil is from November to March and the fruit ripens up to 13 months later. The ana variety is made up of yellow, green and red cultivars, and the green ana is the most popular for water consumption, whether fresh or industrialized (LIRA, 2010).

According to Loiola (2009), the Anao coconut palm is widely used for drinking coconut water, as it has a higher sensory quality than other coconut palm cultivars. Figure 2.2 shows a coconut tree of the Ana Verde variety in full production (ARAGÂO et al., 2002).

Figure 2.2 - Detail of a green dwarf coconut palm.

Source: (ARAGÂO et al., 2002).

2.1.4 Fruit quality and ideal harvest point

The coconut (*Cocus nucifera* L.), which comes from the coconut palm, belongs to the Arecaceae or Palmaceae family and is marketed not only in its traditional form, as a mature coconut, but also in natura, as a green coconut. Therefore, technical considerations about post-harvest care need to be discussed so that the quality of coconut water, both for direct consumption and for industrialization, is as good as possible (LEBER; FARIA 2003).

According to Benassi (2006), the ideal point for harvesting green coconuts is when the water has developed all the sensory characteristics that make it suitable for consumption. The harvest point is determined by associating morphological indicators related to the age or size of the fruit.

Coconut harvesting varies according to the purpose of the plantation. For those intended for the production of unripe fruit, i.e. water consumption, they can be harvested earlier and at a more advanced stage of ripeness, when the purpose is to use the purchase in the industry (FONTES, 2003).

The maturity of the fruit at the time of harvest directly affects the post-harvest quality of the product. The exact moment of harvest directly affects both the quality and the post-harvest useful life of the fruit (WANG, 1997).

Harvesting is related to the degree of maturity of the product in question. Maturity indices are measurements that can be used to determine whether a particular fruit or vegetable is ripe or optimal for consumption. It characterizes the stage of development that allows the minimum acceptable quality for final consumption, taking into account its importance in the marketing chain. In the case of coconuts, it should be noted that they must remain on the plant until they are fully ripe, since their quality can be compromised if harvesting is strictly anticipated (KADER et al.,

1985).

The post-harvest quality of coconuts is highly influenced by various pre-harvest factors, such as temperature in the growing season, light, rainfall, irrigation, fertilization and phytosanitary control. Climatic factors also have a major influence on the quality and nutritional value of the fruit (WESTON; BARTH, 1997).

Quality characteristics develop during the fruit's growth and ripening phases, involving tissue formation, physico-chemical and sensory changes. The main factor influencing the chemical composition of coconut water is the degree of ripeness, the variety, the location of the region and the time of year, which are also determinants of quality (ARAGAO et al., 2002).

According to Aragao et al. (2002), for fresh water consumption, green coconuts should be: free of cracks or mechanical damage; fresh in appearance, free of damage caused by diseases, pests or very low temperatures; free of stains; free of unpleasant odors and tastes; with the maximum volume of water in the central cavity and with sensory characteristics that make it suitable for consumption.

The fruit must be picked with the utmost care to avoid mechanical damage caused by falling. The coconuts are picked by hand and lowered into baskets or bags attached to a rope to prevent them from breaking when they fall. The kernel is thin and delicate, and the cavity is entirely filled with fresh, sugary water, which is highly nutritious and refreshing. The Anao coconut tree, being small in size, makes it easy to harvest the fruit (CARVALHO, 2005).

The harvest point is determined by associating morphological indicators related to the age or size of the fruit, or by counting the leaves on the plant and the presence of certain substances in the water. It can also be determined using the chronological method, i.e. the time elapsed since the opening of the upper spathe (arrow), when the fruit has the maximum volume of water and the best quality (CHITARRA, 2001).

Knowledge of the behavior of sugar levels in water during development is of fundamental importance, as an auxiliary method for determining the best time for the fruit to be harvested and, consequently, for obtaining high quality fruit, i.e. when targeting the water market (MACIEL, 2008).

2.1.5 Green Coconut Market

Brazil is currently experiencing an upward trend in the cultivation of dwarf coconut palms, which are distributed practically throughout the country, as can be seen in Table 2.1. Although coconut cultivation is being stimulated and introduced in various regions of the country, the largest plantations and productions are concentrated in the coastal strip of the Northeast and part of the Northern region of Brazil, favored by tropical climatic conditions, both regions holding approximately 75% of Brazilian coconut production (EMBRAPA, 2011).

Table 2.1- Area planted and production of coconut palms in the regions of Brazil in 2009.

Regions of Brazil	Planted Area ( )^{ha}	Production (thousand fruits)
North East	228.911	1.337.358
North	30.353	281.746
South East	21.564	311.143
Center West	3.934	41.116
South	189	2.003
Total/Brazil	284.951	1.973.366

Source: IBGE Agricultural Production.

Of the ten largest coconut producing states in Brazil, seven are in the Northeast. According to Table 2.2, the state of Bahia leads production, followed by Sergipe and Cearà. These states together account for more than 50% of Brazil's coconut production.

Table 2.2 - Main Brazilian states in terms of planted area, production and productivity of coconut palms in 2009.

Regions of Brazil	Yield (thousand fruits)	Planted Area (ha)	Productivity (thousand fruits/ha)
Bahia	467.080	79.596	5,81
Sergipe	279.203	42.000	6,64
Cearà	259.368	43.448	5,97
Parà	248.188	24.663	10,10
Holy Spirit	157.590	10.625	14,83
Pernambuco	129.822	14.237	9,11
Rio de Janeiro	78.419	4.843	16,19
Paraiba	63.765	11.556	5,52
RioGrandedo North	61.004	21.923	2,78

| Alagoas | 53.083 | 12.524 | 4,24 |

Source: IBGE: Agricultural Production - 2009. Aracaju/SE, 2011.

Bahia is the country's largest coconut producer with 467,080. Paraiba is the 8th largest coconut producer in the country, producing more than 63 million fruits. The superiority of the northeastern states is also significant in terms of the area planted. The three largest producers accounted for the largest percentage (60%) of the total area planted with coconuts in Brazil in 2009 (IBGE, 2009).

In Ceará, the country's third largest producer, more than 47,000 hectares of coconut groves are cultivated, most of which are irrigated and used to produce water both to meet national demand, which stands at around 350 million liters a year, and to meet the growing demand from the international market, which is growing at a rate of 20% a year (NORDESTE RURAL, 2009).

According to IBGE data (2009), Sousa is the largest coconut producer in the state of Paraiba, contributing more than 26 million units of fruit in 2009, in a total of 1315 ha of planted area, followed by Lucena with around 8 million fruits, in an area of 1500 ha. According to the Cooperativa de Agropecuària de Sao Gonçalo (Sao Gonçalo Agricultural Cooperative), the crop generates an annual turnover of around 13 million euros in the municipality of Sousa, employing hundreds of people directly and indirectly. Approximately 35% of this production goes to industries that process dried coconut and a large part to companies and industries that bottle coconut water in Sousa and Ceará. The rest is sold on the domestic market. [a]Production is still small considering that this quantity represents approximately 2% of national production, placing Sousa in 15th place among producers.

Brazil is the fourth largest coconut producer in the world, as can be seen in Figure 2.3, accounting for 30,338,330 tons of fruit in a harvested area of around 28,860 ha, second only to Indonesia with a production of 16.3 million tons of fruit, the Philippines with 14.8 million tons and India with 9.5 million tons of fruit, according to the ranking of the main coconut producers worldwide (FAO, 2007).

Figure 2.3 - Participation of the main coconut producers worldwide

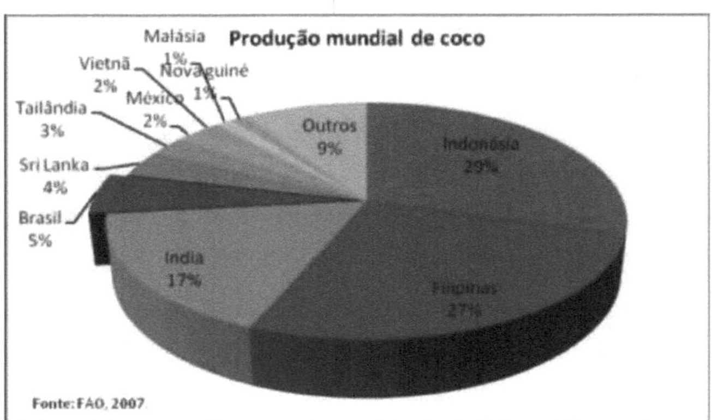

In Brazil, the production and harvested area of green coconuts have grown over the last decade. In 1999, green coconut production was 1.2 billion fruits in a harvested area of 250,000 hectares, in 2009 production was around 1.7 billion fruits in a harvested area of 266,000 hectares, and between 2002 and 2006 production remained stable (FAO, 2007).

Coconut is the third most cultivated fruit in the country, after oranges and bananas. But per capita consumption of coconut water is still very low. The average annual volume consumed by Brazilians (130 mL) is equal to that of whisky, according to a survey published by Amacoco, the country's largest producer of sterilized coconut water. According to estimates by the Brazilian Coconut Association (Abrascoco), coconut water consumption in Brazil in 2003 was 140 million liters/year, with prospects for an increase to 500 million liters/year. Despite this high consumption, it represents only 1.4% of the annual consumption of soft drinks in the country (FOLHA ONLINE, 2004).

In Brazil, the coconut palm is one of the most important perennial crops, specifically in the Northeast region, as it generates many jobs, both direct and indirect, along the production and marketing chain. Therefore, the coconut market can be seen from two points of view: the dried coconut market, in which the fruit is destined for the agro-industry of the Northeast, South and Southeast regions; and another part of the market in the Northeast, in small factories that bottle coconut water from unripe fruit for consumption as fresh water. In general, small producers make up the largest share of coconut production (85%), selling their produce through middlemen (intermediaries and outsourced to the industry), while large coconut producers are the agro-industries themselves, or sell their produce directly to the processing industries (CUENCA et al., 2002).

2.1.6 Coconut water

The formation of coconut water is an eco-physiological strategy of the coconut palm in order to store nutritional reserves to be used naturally as a survival mechanism for the species, to nourish the embryo when the seeds germinate or the seedling during periods of possible environmental stress. Taking advantage of this renewable natural resource, man has increasingly used coconut water in human food and nutrition, in medicine, such as in the case of dehydrated patients or physically exhausted athletes, to replenish potassium, and in biotechnology, in the preservation of goat, sheep, pig and poultry semen, in the induction of cell differentiation, among other applications (ARAGÂO, 2005).

Coconut water begins to form two months after fertilization and undergoes changes in its composition during the development of the fruit, which can be influenced by factors such as: temperature in the growing season, intensity of solar radiation, rainfall, soil class and cultivation, as well as the degree of ripeness, variety of fruit, region and time of year, which also influence the physicochemical and sensory characteristics of the product, as shown in Table 2.3 shown below with data on the physico-chemical characterization of 7-month-old green coconut water. This variability is typical of the consumption of the fruit "in natura" (MACIEL et al., 2009).

Table 2.3- Physical and chemical characterization of green coconut water at 7 months of age.

Parameters analyzed	Values found
Sucrose (mg/100 mL)	280
Glucose (mg/100 mL)	2378
Fructose (mg/100 mL)	2400
P (mg/100g)	7,40
Ca (mg/100g)	17,10
Na (mg/100g)	7,05
Mg (mg/100g)	4,77
Mn (mg/100g)	0,52
Fe (mg/100g)	0,04
K (mg/100g)	156,86
Acidity (%V/P)	1,11
pH	4,91
Total solids (g/100g)	5,84

°Brix	5,00
Vitamin C (mg/100 mL)	1,2
Total carbohydrates (g/100g)	3,46
Protein (mg/100g)	370
Caloric value (Cal/100g)	27,51

Source: (MACIEL et al., 2009).

The basic composition of coconut water consists of 93% water, 5% sugars, as well as proteins, vitamins and mineral salts (ARAGÂO, 2000), with sugars and minerals being its main chemical constituents, while the less important ones are fat, nitrogenous substances, organic acids and dissolved gases (JAYALEKSHMY et al., 1988).

The Ministry of Agriculture, Livestock and Supply, through Normative Instruction No. 27 of July 22, 2009, approved the technical regulations for setting the standard of quality and identity of coconut water. Defining it as a drink obtained from the liquid part of the fruit of the coconut palm (*Cocos nucifera L.*), by means of an appropriate undiluted and unfermented technological process: Coconut water "in natura" - cannot be subjected to any physical or chemical process and is intended for immediate consumption (immediately after extraction). Cooled coconut water - must be bottled as soon as it has been extracted and must be subjected to an appropriate cooling process. It must be kept and marketed under cooling conditions, at a maximum temperature of positive seven degrees Celsius. Frozen coconut water - which has undergone a freezing process, whether or not it has been pasteurized. It must be kept and marketed under freezing conditions with a temperature of minus ten degrees Celsius. Sterilized coconut water - has undergone an appropriate sterilization process and can be sold at room temperature. Concentrated coconut water - which has been subjected to a concentration process and whose measured soluble solids content is equal to or greater than twice its natural concentration. Dehydrated coconut water - which has undergone a dehydration process and has a moisture content of 3% or less. Reconstituted coconut water - which has undergone an appropriate rehydration process. All green coconut water sold must have sensory characteristics such as appearance, color, characteristic odor, and be within the physicochemical parameters of fixed acidity in citric acid (g/100 mL), pH of at least 4.3 and soluble solids (in °Brix at 20 °C) of at most 7.0 (BRASIL, 2009).

Green coconut water can be consumed both fresh and processed, and its useful life depends on the preservation methods applied. These methods aim to inhibit enzymatic action and guarantee the microbiological stability of coconut water after the fruit has been opened, while maintaining its original characteristics as much as possible.

Traditionally, coconut water is sold inside the fruit itself, a practice which involves problems related to transportation, storage and the perishability of the product. In order to allow it to be consumed in places outside the producing regions, its industrialization is of fundamental importance, with the aim of reducing the volume and weight transported and, consequently, reducing transport costs, as well as increasing its shelf life (CUENCA, 2007).

Coconut water is a sterile product when inside the fruit; however, due to its nutrient-rich composition, it is very susceptible to microbial growth, making microbiological control necessary. Another important factor is the enzymatic activity naturally present in the liquid. Although these enzymes have specific and vital purposes for the fruit *in vivo*, when they come into contact with the atmosphere they trigger undesirable reactions such as the development of a pinkish color (ANDRADE, 2008).

Coconut water should be consumed within a maximum of ten days after harvesting; from then on, deterioration processes begin which mainly compromise the acidity of the liquid (Aragao et al., 2001). It is extremely perishable and this characteristic is directly related to the conditions to which the fruit is exposed during harvest, post-harvest and marketing. High temperatures, mechanical damage, improper handling and inappropriate storage conditions accelerate the process of water deterioration, reducing its useful life (RESENDE et al., 2008).

Coconut water plays an important role in the ripening and germination of the fruit and its composition varies markedly during the ripening process. It can represent a rival product to sports drinks due to its ability to replenish electrolytes, serving as a basis for adding value to coconut products, with vast commercial potential, due to its nutritional value, being sterile, being a natural drink containing a good amount of minerals, with a mild aroma and taste and consumed by all ages (AGRICULTURE, 2003).

According to Chang: Wu (2011), coconut water is a drink that is gaining popularity in the beverage industry due to its high nutritional value and some potential therapeutic properties. Because it is useful in preventing many health problems, including dehydration, digestive problems, fatigue, heatstroke, diarrhea, urinary tract infections and also reports that the presence of catechin, epicatechin, present in natural coconut water, have antimicrobial, antioxidant and anticancer properties, also rich in cytokinins, a hormone capable of delaying the appearance of aging characteristics in human skin.

2.1.7 Conservation methods for coconut water

Whether green coconut water can be safely consumed in its fresh or processed form depends on the preservation methods applied. There are basically two methods of preserving green coconut water:

bottled and refrigerated. The basic difference between the two is that in one of them, auxiliary treatments can be applied (formulation and pasteurization) which can increase the product's shelf life, allowing for greater flexibility in marketing the product. For the product without auxiliary treatments, the shelf life is around three days. After this period, both the microbial load can increase and biochemical reactions can trigger color alteration processes. In any form of preservation, process time should be optimized and exposure to air minimized. The stages of the process must be rigorously followed and monitored to guarantee quality from a health and consumer safety point of view (ABREU, 2005).

By applying auxiliary treatments, the product's shelf life can be extended to up to six months. In the formulation stage, additives are chosen to perform specific functions (preservative, antioxidant, acidulant). The pH must be corrected with suitable organic acidulants and kept below 4.5 (ABREU, 2005).

Refrigerated coconut water is sold in PET (polyethylene-terephthalate) plastic bottles, as well as in cups with heat-sealable lids or low-density polyethylene (LDPE) bottles. There are manual bottle fillers as well as automated systems. The filling stage should be carried out in the shortest possible time, preferably with the product pre-cooled. The storage temperature should be kept at around 5 to 8 °C (ABREU, 2005). Figure 2.4 shows the stages involved in the various forms of product preservation proposed by (ABREU, 1999).

Figure 2.4- Flowchart of the stages involved in the coconut water preservation process.

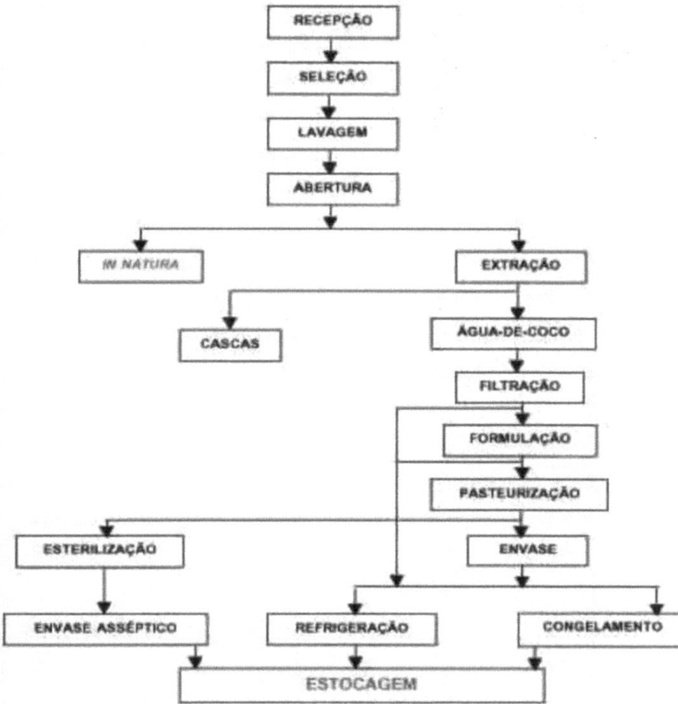

Source: (ABREU, 1999).

2.1.8 Description of the stages in the coconut water bottling process

Regardless of the preservation method used, some steps are common to any coconut water packaging process. It must be taken from healthy, clean fruit, free from parasites and animal or plant debris. It must not contain fragments of inedible parts of the fruit or substances foreign to its original composition. This product must be obtained through an appropriate technological process that maintains its natural characteristics. Figure 2.5 shows the main stages involved in the process of preserving green coconut water.

Figure 2.5 - Stages in the production process of refrigerated green coconut water.

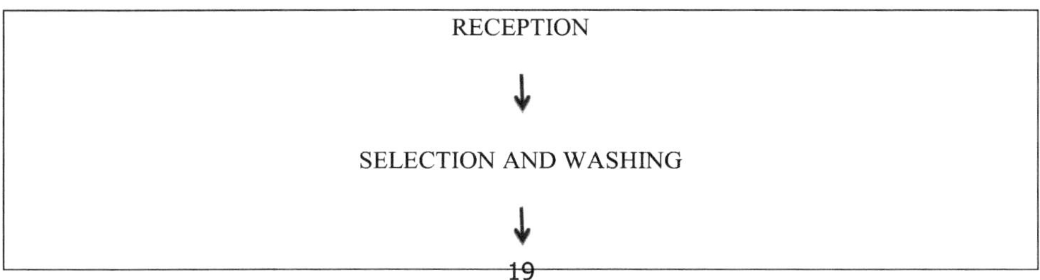

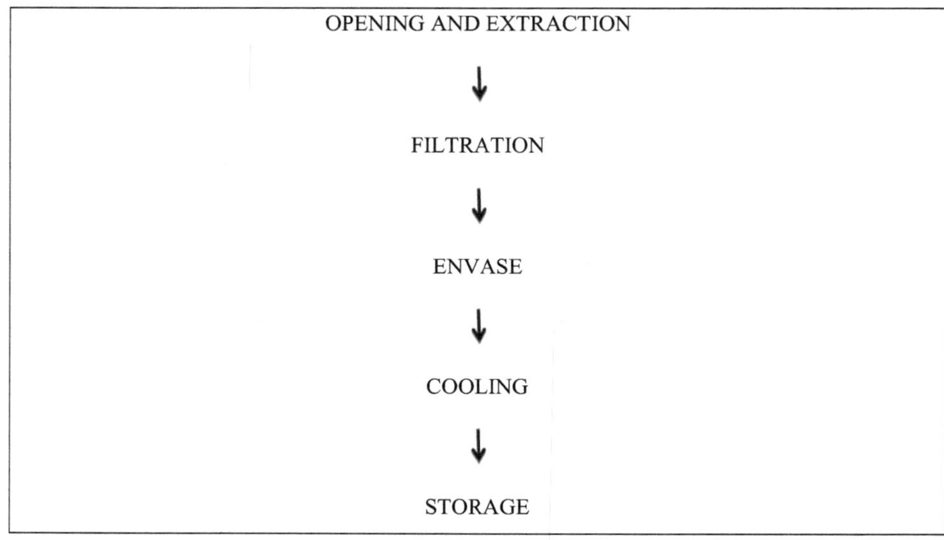

Source: (ABREU, 1999).

Receiving and weighing

The fruit should be received in a reserved place, preferably a platform, where the trucks or other means of transport can pull up, making unloading easier. When they arrive at the industry, the fruit must be unloaded and weighed. The raw material must be counted and then a representative sample taken from the load to carry out the initial analysis (volume of water in the cavity, °Brix and pH) to check its quality. It is important to characterize the batches, especially when they come from different suppliers. This will make it possible to standardize the final product as much as possible, since the composition of the water depends on the variety and the ripeness of the fruit. The coconuts can arrive at the production area loose or preferably in bunches. The reception area should always be sanitized, washed with water and detergent and then with chlorinated water (100 ppm free chlorine). Accumulated waste should be collected daily.

The fruit can be stored for some time, but they should be kept in a cool, ventilated place (20 °C), protected from the sun and wind. This environment is suitable for storing coconuts for 15 days when kept in bunches and, when stored under refrigeration (12 °C), they can be stored for up to 30 days. Outside the bunches, they last around 10 days.

Selection and washing

Coconuts to be processed must be visually sorted. Diseases in coconut trees can cause spots and lesions on the fruit, which will allow microorganisms to enter, which is why all unsuitable fruit should be discarded.

Selection must be careful and done manually by trained employees. Uniformity in the ripeness and variety of the fruit is recommended. This stage is very important, as choosing overripe or damaged coconuts can compromise the quality of the final product.

Proper sanitization of fruit is essential to guarantee the health quality of the product and its best preservation.

Washing can be done manually (immersion in stainless steel, PVC or tiled tanks) or automatically (system with conveyor belts, brushes and sprinklers), and must be carried out in three stages: Pre-washing with treated water, washing with chlorinated water, rinsing. After the washing procedure, the coconuts should be kept in a clean, dry place until all the water has drained out.

Opening and extracting water

During processing, the coconut opening phase is considered a critical point, since a slow opening system compromises the speed of the enzymatic process, allowing undesirable reactions to occur in the product. Manual opening, being a slow process, can allow unwanted reactions to occur in the water. In this sense, it is recommended to minimize the time the coconut water is exposed to air.

The coconut is pierced using manual equipment, which can be a drill or a knife, always made of stainless steel. When piercing the coconut, it is recommended to insert the instrument into the upper part of the fruit, where the pedicel (the part that holds the fruit in the bunch) is located. The hole for removing the water from the fruit should be very wide, making it easier to drain. The contact between the water and the fibrous part of the fruit, in the presence of oxygen, can also cause enzymatic reactions that can alter the product's characteristics.

All utensils and equipment should be made of inert material (Teflon, nylon, polyethylene, glass) and those for cutting and handling the fruit should be made of stainless steel. Accumulated garbage (coconut shells) should be removed from the area periodically to avoid sources of contamination. Most of the opening and extraction systems are developed by the fillers themselves to meet their specific needs.

Filtration

After opening, the water must be poured into a trough or synthetic collector with meshes of 60 to 100 threads /2 cm or a sieve, with fine meshes, less than 0.3 mm open, used to retain the solids or residues from the opening stage. As these solids are usually residues of the shell, they have a reasonable proportion of components that can stimulate the action of enzymes that alter the color of coconut water, making it a pinkish or darker product, or even modify the taste. It is important that the filtered coconut water is taken to a stainless steel tank to pre-cool the product, minimizing the risk of contamination. Coconut water can also be filtered through plastic or stainless steel filters and the liquid collected in tanks made from the same material as the filters.

Packaging

The filtered and pre-cooled coconut water should be pumped into a dosing machine, which should be set to fill the container in pre-defined quantities. A refrigerated stainless steel tank with a filling tap can also be used. In this type of process, the most commonly used packaging is plastic bottles or 300 ml cups made of PET (polyethylene terephthalate) or polyethylene. Most of the time, the bottles are closed by hand, but a cap sealing machine can also be attached to ensure sealing, and hot air sealers are used for the cups.

Cooling

After filling and sealing, the packages are placed in cold rooms or refrigerators. This should be done as quickly as possible to preserve the original characteristics of the water and prevent microbiological growth or changes in color.

Storage

Green coconut water should be kept refrigerated until consumption. The recommended temperature for storage in cold rooms or refrigerators ranges from 6 °C to 10 °C. The amount of product inside the chamber or refrigerator must be observed, which should not be excessive, in order to allow good air circulation between the walls of its compartments and between the packages. The basic rule for the movement of stored stock must be observed in terms of the order in which the goods enter and leave, due to the expiry date.

The shelf life of coconut water refrigerated at 6 °C is around 3 days, depending on the hygiene conditions used in processing, the time elapsed between opening the coconut and filling it, and the degree of ripeness of the fruit used as raw material. It is important that the cold chain is not broken throughout the distribution and sale of coconut water, until it is consumed, in order to guarantee the quality of the product.

Label

The following information must appear on the packaging label: name: green coconut water; quantity in ml; date of manufacture; expiry date. Expressions: natural (provided no additives have been added) contains: naturally mineral salts (when supplementary nutritional information is used), or similar expressions. Mandatory statement: Once opened, keep in the refrigerator and consume within, or similar expressions. Name: Brazilian Industry and Ministry of Agriculture, Livestock and

Supply. Name and address of company, CGC and state registration (EMBRAPA, 2005).

2.1.9 Industrialization and marketing of coconut water

The consumption of green coconut water in places far from its region of production has always been problematic, as it depends on the logistics of the fresh fruit to these places, which compromises the quality of the product due to the various contaminations that can occur during its transportation under high temperatures and prolonged periods of time (CARNEIRO, 2007).

According to Gaiva et al (2004), the importance of the coconut palm is due to the products and by-products destined for industry, as well as the production of fruit for consumption as fresh or bottled water. The socio-economic importance of this palm is due to the diversity of products obtained from the different parts of the plant.

The marketing of green coconut water transported inside the fruit itself involves cost increases related mainly to transportation, storage and the perishability of the product. In this sense, the extraction and packaging of green coconut water appears to be an alternative capable of eliminating the risks of microbiological contamination, nutritional variations, sensory alterations and changes in the color of the final product. This technology involves a set of simple steps designed to preserve the microbiological and sensory quality of coconut water after it has been extracted. It basically consists of maintaining the temperature of the product at low levels so that the rate of enzymatic reactions and microbial growth are minimized. As well as providing and adopting Good Manufacturing Practices (GMP) to guarantee product quality. This technology has the quality of maintaining the natural flavor of coconut water. It differs from pasteurization technology, which has sensory deficiencies (CRIBB, 2006).

The main impacts of coconut water bottling technology have been to add value, formalize the trade in green coconuts, increase the product's shelf life, optimize the use of the fruit, reduce the participation of intermediaries who add to the final cost of the product and generate jobs in the agro-industrial sector (CRIBB, 2008).

The aim of industrializing coconut water is to obtain a product that preserves its natural characteristics as much as possible, extending its useful life and making it easier to consume outside the regions where it is grown (MARQUES; GALLI, 2007).

2.1.10 Refrigerated storage

In the case of refrigeration, temperatures higher than the product's freezing point are adopted. The temperature range varies according to the type of food, time and storage conditions. In general, refrigeration can be used for basic food preservation or complementary preservation (ASSIS, 2012).

According to Garbutt (1997), storing perishable foods in refrigeration guarantees their preservation for periods of time that can range from a few days to several weeks, which will be reflected in a reduction in the growth rate of the microorganisms present in the food and consequently an increase in its useful life. There are different ways of extracting, preserving and packaging coconut water. Innovations in packaging and marketing methods have made it easier for consumers to find coconut water in all seasons of the year, to transport it easily without the need, in some cases, for refrigeration and to store it in less space than the fruit itself Frasseti et al., (2000). The industries that process green coconuts to obtain water have faced problems of enzymatic and/or microbiological origin that change the sensory characteristics and potability of the product (HOFFMANN et al., 2002).

Refrigeration is still the most economical method for long-term storage of fresh fruit and vegetables, while other storage methods are used to complement refrigeration, such as controlling or modifying the atmosphere (SANTOS FILHA, 2006).

Temperature is one of the most important factors in food preservation, preparation and cooking. The most prominent aspects related to foodstuffs and temperature control are described in the Regulation on the Hygiene of Foodstuffs (EC Regulation 852/2004). The most critical factor in guaranteeing the quality and prolonging the life of food is temperature. Temperature abuses have a particularly adverse effect, causing easy spoilage (Pais, 2007). The ideal storage conditions vary widely depending on the product and correspond to the conditions under which it can be stored for as long as possible without appreciable loss of its quality attributes, such as: taste, aroma, texture, color and moisture content. The storage period depends above all on the product's respiratory activity, susceptibility to moisture loss and resistance to disease-causing microorganisms (CHITARRA; CHITARRA, 2005).

According to Kader (1992), comparing the use of refrigeration and storage at room temperature in mangoes (*Mangifera indica* L.) stored at 15 °C, he detected less degradation in ascorbic acid and chlorophyll and higher values in total acidity. In pineapples (Annonasquamosa L.), the post-harvest shelf life can be increased by reducing the storage temperature (TSAY, 1988), with the limit for refrigeration being around 15 °C (BROUGHTON; TAN, 1979). Hoffmann et al. (1994) obtained satisfactory results when they used refrigeration to preserve guava fruits (*psidiumguajava* L.) "Serana".

According to Garcia (1980), green coconuts should be stored in a temperature range of 0 °C to 1.7 °C and relative humidity between 80 and 85%, and under these conditions they can be stored satisfactorily for one to two months. However, Assis et al. (2000) and Resende et al. (2001) recommend a temperature of 12 °C for refrigerated storage of green coconuts, suggesting that temperatures lower than this can cause cold damage. Despite the common use of cold for post-harvest preservation of green coconuts, storage can cause alterations in the fruit, depending on the binomial temperature - exposure time of the fruit, as well as its stage of ripeness at the time of storage (POWELL, 1988).

Farias et al. (2006), evaluated changes in the physical characteristics of Anao verde coconut fruits during refrigerated storage at different temperatures, and found that fruits stored at 6 °C did not offer resistance to the cold, however, taking into account the darkening of the skin caused by the cold, the fruits were preserved at 6 °C for 7 days, while at 9 °C for 3 weeks.

Temperature is one of the most important factors in the degradation of plant tissues and determines the speed of biochemical reactions associated with senescence. The quality and freshness of the product are essential elements for marketing and increasing sales, improving the cost-benefit ratio and customer satisfaction (TERUEL, 2008).

Refrigerated storage is one of the most important methods used to extend the useful life of fruit, vegetables and products derived from them, such as juices. When this type of product is kept at temperatures below a certain critical limit, or above freezing temperature, it results in quantitative and qualitative post-harvest losses (SOARES, 2011).

Cold preservation has the advantage of preserving a large part of the nutritional and organoleptic value of food. However, it has the disadvantage of not eliminating the microorganisms present in the food, nor the harmful action of their toxins, but only inactivating them, so that when they find favorable environmental conditions they resume their activity. It is therefore important to guarantee the good quality of the raw materials before chilling and freezing, as well as carefully controlling the temperature during these processes (INOVADORA 2009).

Refrigeration is a process that brings tangible benefits, even more so in countries with a tropical climate, such as Brazil. Particularly as it is a major producer of fruit and vegetables and still a small exporter, with great market potential, provided aspects related to post-harvest, quality standards and marketing are met (TERUEL, 2008). Likar and Jevsnik (2006) established that the cold chain for perishable foods should be regulated at temperatures $\leq$ 5 °C.

2.1.11 Physical, chemical and physico-chemical characteristics.

Knowing the chemical and physical-chemical composition of fruit has been the subject of research over the years and is essential if its technological use is to be optimized (OLIVEIRA et al., 2006). During the development of the coconut, the water undergoes a series of important changes in its physical, chemical and chemical characteristics that can significantly influence its quality and acceptance (ARAGÂO et al.,2002).

The water in the fruit of the Anao verde coconut palm is extremely perishable due to its physical, chemical and biochemical characteristics. This perishability is directly related to the conditions in which the fruit is exposed during harvest, post-harvest and marketing. Thus, high temperatures, mechanical damage, handling and inadequate storage conditions accelerate the process of water deterioration, altering its taste, nutritional quality and reducing its useful life (RESENDE et al., 2007).

The concentration of hydrogen ions in the pH of a food is important because of the influence it has on the types of microorganisms that are able to multiply and, therefore, on the changes they should logically produce (GAVA, 1998).

According to Machado (2008), several factors make it essential to determine the pH of a foodstuff. Measuring pH is important for determining food spoilage with the growth of microorganisms, enzyme activity, texture, retaining the taste and smell of fruit products, checking the ripeness of fruit and choosing packaging. Influence on palatability, definition of the heat treatment temperature to be used, selection of the type of cleaning and disinfection material, definition of the equipment with which the industry will work and selection of additives (CHAVES, 1993).

According to Souza et al. (2002), the pH of water varies little throughout development, but changes with ripening, increasing as it goes on. It is around 5.1 to 5.2 in the water of ripe fruit, and 4.7 to 4.8 in green coconuts (JAYALEKSHMY et al., 1986; SANTOSO et al., 1996).

According to Pinheiro et al. (1984), the content of soluble solids is of great importance in fruit, whether for fresh consumption or for industrial processing, since high levels of these constituents in the raw material mean less sugar is added, water evaporation takes less time, less energy is used and the product yields are higher, resulting in greater savings in processing.

As the name suggests, total soluble solids (TSS) represent all the soluble constituents of the fruit, mainly sugars, organic acids and salts in a given solvent, which in the case of food is water; sugars make up the bulk of TSS. In general, as ripening progresses, the TSS content evolves, but as it is strongly influenced by climatic conditions and handling, it has been used more as a quality parameter than the actual point of harvest (MARTINS, 2005).

Santos Filha (2006) reports that sweetness is linked to the proportion of total sugars and acidity, and these vary considerably with the cultivar, soil class and climatic conditions.

Nery et al. (2002), evaluating Anao coconut varieties in the state of Parà, observed that the TSS values found were well below those expected for coconut water. Many factors can influence this characteristic, from climate to cultural management and the origin of the plant material.

In a study carried out with dwarf coconuts, Souza et al. (2002) concluded that the maximum content of total soluble solids in coconut water is reached in the 7th month, the content of organic acids decreases and that of total and reducing sugars increases. In this same variety, Nery Bezerra; Lobato (2002), found average contents of 4.32 °Brix in the product in the 7th month of ripening.

Kluge et al. (2002) report that the acid content of a fruit is given by its total titratable acidity (TTA). Acidity can be used, together with sweetness, as a reference point for the degree of ripeness. For some fruits, determining the point of harvest by determining the total titratable acidity is unreliable, due to the fact that there is little variation in this characteristic throughout ripening.

The TSS/ATT ratio is one of the most widely used ways of evaluating flavor and is more representative than measuring sugars or acidity alone. This ratio gives a good idea of the balance between these two components, and the minimum solids content and maximum acidity content should be specified in order to get a better idea of the flavor (CHITARRA; CHITARRA, 2005). According to Santos Filha (2006), increased acidity in coconut water makes it less palatable.

Vitamin C is the most important nutritional component to be determined, characterized by its antioxidant character and for being a catalyst in biochemical reactions involving hydroxylation. It plays a fundamental role in human nutrition and as it is the most thermolabile vitamin, its presence indicates that the other nutrients are probably also being preserved in the food (CHITARRA; CHITARRA, 2005). Aroucha; Carlos (2000) and Vianna et al. (2008) found vitamin C values in dwarf coconut water ranging from 0.95 mg/100mL in the 4th month to 3.15 mg/100mL in the 12th month. However, the sixth month is the richest in vitamin C. The problem is that during this period the acidity is higher and the taste is impaired. As for minerals, potassium always appears in good proportions.

The composition of ash does not correspond to the amount of mineral substances present in the food, due to losses through volatilization or even reaction between other components. Ash is considered a general measure of quality and is often used as a criterion for identifying foods. Ash contains mineral components, including calcium, magnesium, iron, phosphorus and lead. A very high ash content indicates the presence of adulterants (OLIVEIRA et al., 1999). Coconut water is rich in minerals, regardless of the age of the fruit, but there are variations with age, planting regions and cultural treatments (PENHA et al., 2005). Santoso et al. (1996), when analyzing the constituents of green dwarf coconut water, found an approximate composition of 5.55% dry matter for the fruit at the 12th month of ripeness. The ash, crude protein, total lipid and carbohydrate contents as a percentage of dry matter were 8.42%, 9.36%, 2.67% and 79.5%, respectively. Jayalekshmy et al. (1988) obtained 2.0% total sugars, 0.076% protein, 0.083% lipids and 0.54% ash.

The least present constituents are fats and nitrogenous substances Maciel, (2008). According to

Srebernich (1998), the protein and fat content of coconut water increases with the age of the fruit and is dependent on variations between varieties and harvests.

Lipids are made up of carbon, hydrogen and oxygen and provide 2.23 times more energy/kg when oxidized than carbohydrates. Fats serve as energy suppliers and are broken down in cells during cellular respiration. Lipids are also a source of essential fatty acids for the human body and serve as transporters of nutrients and fat-soluble vitamins, such as A, D, E and K (PINHEIRO et al., 2005).In water from ana variety coconuts, Fagundes Neto et al. (1989) observed a drop in lipid content from the 6th to the 9th month (277.29 to 57.87%), and then an increase, reaching 135.42% in the 12th month.

As well as sugar, coconut water contains protein (around 370 mg/100mL), vitamins (ascorbic acid, nicotinic acid, biotin, riboflavin and folic acid) and minerals such as Na, Ca, Fe, K, Mg and P (ROSA; ABREU, 2000).

Coconut water contains a small amount of protein. The percentage of alanine, arginine, cystine and serine in young coconut water is higher than in cow's milk. Green coconut water does not contain any protein complexes that could cause shock in patients (MARQUES, 1976).

Compared to green coconuts at the 6th month, water from mature coconuts has higher concentrations of crude protein and total lipids. However, the levels of ash and carbohydrates are lower (SANTOSO et al., 1996).

Fagundes Neto et al. (1993) reported that coconut water of the ana variety showed a drop in lipid content from the 6th to the 9th month (277.29 to 57.87%), and then an increase, reaching 135.42% in the 12th month.

Sugars can be quantified by chemical methods. Due to the respiratory process, in which sugars are oxidized to produce energy, the concentration of these compounds changes progressively in plant cells, and represents a parameter that can be used to monitor post-harvest conditions, in conjunction with other evaluations (CHITARRA; CHITARRA, 2005).

According to Narayan et al. (2000), coconut water contains mainly glucose, fructose and sucrose. The fruit of the coconut palm can reach full maturity between 11 and 12 months after flowering, and at approximately seven months (220 days) the sugar concentration in the water reaches its maximum, which is when it reaches its best quality.

Jaylekshmy et al. (1986), when reporting chemical changes in coconut water at different stages of development, stated that reducing sugars (glucose and fructose) decreased from 4% to 0.2% and that sucrose reached around 90% of total sugars. Similar behavior was found by Santos et al. (1996), Srebernich (1998) and Souza et al. (2002) in Ana coconut fruits, where in the eighth month after fertilization, reducing sugars predominated over non-reducing sugars and, from the ninth month onwards, the situation was reversed, with a consequent loss of water quality.

As the fruit ripens, the content of reducing sugars decreases and the content of non-reducing sugars increases. Thus, glucose and fructose predominate in green coconut water, with sucrose playing a lesser role. In the water of mature coconuts, the concentrations of glucose and fructose are also higher, but sucrose plays a greater role (JAYALEKSHMY et al., 1988; SANTOS et al., 1996; SREBERNICH et al., 2000).

Reducing sugars predominate over non-reducing sugars until the 8th month of development, reversing this situation from the 9th month onwards, with a sharp decrease in reducing sugar content until the end of fruit ripening and consequent loss of flavor (SOUZA et al., 2002).

Coconuts undergo a series of stresses when they are harvested, due to changes in their environment. Therefore, Chitarra and Chitarra (2005) consider biochemical transformations to be the main events responsible for changes in sensory attributes and shelf life.

Turbidity is the transparency of water due to suspended matter. It is assessed by measuring the amount of light reflected, giving the magnitude of the solids suspended in the sample, but cannot be immediately associated with the amount of suspended solids; Santos filha, (2006). Turbidity is associated with the color of the water, and in general, an increase in suspended solids changes the color of the water. The Brazilian legislation Brasil (2009), used for the industrialization of coconut water, states that the appearance can vary from transparent to translucent, defining that the transparent appearance is when the light passes through the liquid, allowing the objects on the other side to be seen, and the translucent appearance is when the light passes through the liquid, without allowing the objects on the other side to be seen, in addition to the fact that the presence of a small amount of coconut pulp particles does not disqualify the product.

Coconut water shows significant changes in color during the development phase, which is a parameter that must be taken into account when establishing its quality (MACIEL, 2008). According to Chitarra and Chitarra (2005), color is more directly related to the consumer's perception of appearance, while the concentration of pigments may be more directly related to the maturity of the product. It is of interest that the product shows intensity and uniformity of color, which can be assessed in the peel and pulp of fruit and vegetables using different methodologies.

Electrical conductivity can also be used as a good tool for characterizing food products (PALANIAPPAN; SASTRY, 1991).

According to Fellows (2006), the physical property that reflects the resistance to flow and results from the internal friction between fluid layers is viscosity. Based on physical parameters, a liquid is considered to be made up of a series of overlapping layers. When it flows over a surface, the upper layer moves faster than the lower ones, dragging them along at a lower speed, and so on. In the food industry, the study of viscosity applied to liquid foods is of fundamental importance. The viscosity of many liquids changes with changes in temperature, which also influences mouthfeel, syrups, yogurt, juices and other foods. It is important in food quality control and the development of new products (texture and consistency).

The coconut normally used for drinking water is between the 5th and 7th month, because until the 4th month of ripeness the fruit is extremely small. It is considered that the 7th month is the best age to harvest the fruit for the production of drinking water, because in addition to the greater quantity of water, at this time it has the best balance of minerals, sugars, pH and acidity. This makes it very tasty and it is rich in fat-free nutrients, which makes it an excellent choice for human health (FAGUNDES NETO et al., 1989).

2.1.12 Microbiological evaluation

The growth of national and international trade and the ease with which people can travel today increases the number of people who consume food and increases the spread of potentially pathogenic agents. Nowadays, the world is increasingly interconnected and interdependent, so an outbreak of food-borne illnesses locally, with relative ease, becomes a potential threat to the whole world.

There are several factors that contribute to the occurrence of FBD, which can be divided into three distinct groups: factors that influence food contamination; factors that allow potentially pathogenic microorganisms to proliferate; and factors that allow pathogenic microorganisms to survive in food. Food-borne illness (FBD) is a syndrome of an infectious or toxic nature caused by the ingestion of food and/or water containing etiological agents of biological, physical or chemical origin in quantities that affect the health of the individual consumer or a group of the population, which can take a variety of forms, ranging from mild ailments to more serious situations that may require hospital care, and can even be fatal (PARANA, 2009).

The most commonly reported contamination factor comes from hand contact with food. Effective hand washing can prevent the transmission of enteric infections (GREIG; LEE, 2009). According to Smith et al, (2004), approximately 76 million ATDs occur in the USA, of which 365,000 lead to hospitalization and 5,000 result in death. And every year thousands of North American consumers suffer from some kind of foodborne illness with symptoms ranging from mild to fatal.

All foods, whether of animal or vegetable origin, are contaminated by the most diverse types of microorganisms, which are part of their usual flora. To maintain the multiplication process, these microorganisms need favorable conditions (substrate quality, nutritional value, temperature, pH, humidity (GERMANO, 2003).

The microbiological quality of food is dependent firstly on the quantity and type of microorganisms initially present (initial contamination) and then on the multiplication of these germs in the food. The quality of the raw materials and hygiene (of environments, handlers and surfaces) represent the initial contamination. The type of food and environmental conditions regulate multiplication (HOFFMANN, et al.; 2002).

The microorganisms present in food can cause harmful chemical alterations, resulting in what is known as microbial spoilage, resulting in alterations to the color, smell, taste, texture and aspects of the food. These changes are the consequence of the metabolic activity of microorganisms that are simply trying to perpetuate their species, using food as a source of energy. Microorganisms can pose a health risk and are known as pathogens, which can affect both humans and animals. They can reach the food through a number of routes, always reflecting poor hygiene conditions during production, storage, distribution or handling at a domestic level. They can also cause beneficial changes in a food, modifying its original characteristics in order to transform it into a new food (FRANCO; LANDGRAF, 2007).

Coconut water is a weakly acidic and sterile solution which still has its initial microbiota in the palm tree, due to the presence of a few viable microorganisms. However, the outer shell of the coconut or seed can be damaged, allowing bacteria to penetrate the white pulp, thus causing contamination during harvesting, storage on the ground (due to possible contact with sand and manure), loading, transportation and unloading (ABREU, 2005).

When coconuts come into contact with soil, foliage and wind, they have a natural microbial load, which can include various pathogens such as *Salmonella*, Listeria monocytogenes, enteropathogenic E. coli, Clostridium botulinum spores, C. perfringes, Bacillus cereus, toxigenic molds, protozoa, among others. Contamination with soil, which is often wet, at the growing sites can lead to truckloads of contaminated products. The control of the stages that precede the consumption of coconut water, which does not undergo a decontamination process, such as cultivation, harvesting, transportation, reception, toileting, sorting and washing, plays a fundamental role in the prevention of biological hazards and, for this reason, some of these stages can be considered Critical Control Points (CCPs), as pointed out by SCHMIDT et al. (2004).

The main groups of microorganisms that are important in determining the healthiness of coconut water for consumption are: The total count of aerobic mesophilic microorganisms is a rough measure of the bacterial content, the conditions of temperature abuse and the sanitation of the process. The fecal coliform group should not be correlated directly with fecal contamination, but its presence can mean poor processing or post-process contamination, or both. The presence of E. coli, also known as "coliforms at 45 °C", indicates direct or indirect fecal contamination, assessing the sanitation of the industry. *Salmonella*, when detected in the product, condemns the batch under analysis. Molds and yeasts in high counts indicate poor sanitation in food processing or poor selection of raw materials introducing contaminated products. They may or may not present health risks (MASSAGUER, 2007).

Mesophilic Aerobes (AM)

Aerobic mesophilic microorganisms show optimal growth between 20°C and 45°C. Their count provides an estimate of total microbial contamination and high counts are usually related to poor quality and reduced shelf life of products Jay (2000), and their detection and enumeration is used both for quality control and for the efficiency of equipment and utensil sanitation practices during product production and processing (FRANCO; LANDGRAF, 2007). Therefore, a high count of aerobic mesophilic bacteria means the occurrence of conditions favorable to their multiplication (SOUZA et al., 2004).

Bacteria from the coliform group

The term coliform includes the bacterium *Escherichia coli* and several species belonging to other genera of the *Enterobacteriaceae* family, capable of fermenting lactose with gas production when incubated at 35 - 37 °C for 48 hours. They are rod-shaped, Gram-negative, non-sporulating bacteria. In natural foods and on the surfaces of utensils and equipment in food industries, various types of bacteria from the *Enterobacteriaceae* family remain longer than *Escherichia coli*. The presence of total coliforms in food does not necessarily indicate recent fecal contamination or the occurrence of enteropathogens, but it does indicate contamination during the manufacturing process or even post-processing contamination (RODRIGUES, 2007).

Escherichia coli belongs to the group of total coliforms that have the ability to continue fermenting lactose with gas production when incubated at temperatures of 44 - 45 °C. The use of Escherichia coli as an indicator of fecal contamination in water was proposed in 1892 by Teobaldo Smith, since this microorganism is found in the intestinal contents of humans and homeothermic animals (FRANCO; LANDGRAF, 2007).

CT and EC counts can be used to estimate hygiene failures and fecal contamination, and high counts of these groups of microorganisms are generally related to significant levels of enteropathogens (JAY et al., 2005).

Most probable number (MPN) technique

The MPN technique is a way of estimating the number of microorganisms present in a sample. It is based on the statistical probability of a certain number of microorganisms being present in the sample when a series of positive results occur. This estimate is obtained by preparing successive decimal dilutions of the sample and transferring aliquots of these dilutions to a series of three or five tubes containing suitable culture medium. In the case of coliform analysis, the positive result is evidenced by the presence of gas inside the inverted tube (Durham tube), contained inside the tube with culture medium. The MPN method is therefore an indirect form of measurement, in contrast to the plating technique which can be considered a direct method (KONEMAN, 2008).

Molds and yeasts (BL)

Molds grow rapidly on food residues that adhere to the surfaces of equipment and contaminate the food that passes through (FRANCO; LANDGRAF, 2007). Various molds, and possibly some yeasts, can pose a danger to human and animal health by producing toxic metabolites, known as mycotoxins, most of which are thermostable compounds that remain active after heat treatment (DILLON, 1998). Yeasts differ from molds in that they are usually and predominantly unicellular. As single cells, yeasts grow and reproduce more quickly than molds. They are more efficient at carrying out chemical changes, due to their greater area/volume ratio. They are easily differentiated from bacteria due to their larger size and morphological properties (JACKSON, 2007).

Molds and yeasts are important indicators of the efficiency of equipment and utensil sanitation practices during food production and processing. Molds and yeasts produce a wide range of metabolites, many of which have been associated with the appearance of pathological effects in animals and humans. The term toxins has been used for these compounds, which include a wide variety of structures, including some that are relatively simple. Mycotoxins occur in the mycelia of filamentous fungi and are produced by a wide variety of species (VERLINDER; NICOLAI, 2000).

The presence of fungal agents in food is undesirable due to their high enzymatic arsenal, which has a great capacity to deteriorate food. These agents are responsible for the development of allergic conditions and/or gastric inflammation, due to the inhalation and ingestion of their spores, respectively (SOUZA, 1997).

Salmonella

Salmonella belongs to the Enterobacteriaceae family, is Gram-negative, produces acid gas from dextrose and does not normally ferment lactose. There are more than 2000 serological types classified according to their antigens, which have varying degrees of virulence in warm-blooded animals. *Salmonella* is widespread in nature, frequently occurring in the intestinal tract and feces of humans and animals. They can be found in companies that process or handle foodstuffs, other sources are domestic animals and rodents (KONEMAN, 2008). It multiplies at temperatures between 7 °C and 49.5 °C, with 37 °C being the optimum temperature for development, at which point contaminated food can become infectious within 4 hours. Below 7 °C, for most serotypes, there is no multiplication (GERMANO; GERMANO, 2003).

International data points to *salmonella* as the main agent of outbreaks in the USA. In that country, between 1993 and 1997, 32610 cases were recorded, with 13 deaths (ALVES FILHO, 2003).

The signs and symptoms of enteric *salmonella* infection appear 12 to 36 hours after eating the contaminated food. The most common symptom is diarrhea, although some people may experience nausea, vomiting, abdominal pain or headache, either as isolated symptoms or in all possible combinations. Sometimes the infection is noticed retrospectively when the patient with few or almost no symptoms develops arthritis two weeks later (CAETANO et al., 2004).

Some researchers report that the increase in *Salmonella* problems is due to various factors, including the increase in the amount of prepared and semi-prepared food, the use of incorrect food storage methods, and the growing habit of consuming raw or incorrectly cooked food (JACKSON, 2007). According to the WHO, salmonellosis is one of the most common public health problems and represents a significant cost in many countries. Millions of human cases are reported worldwide every year, resulting in thousands of deaths (WHO, 2005).

Psychrotrophic

Psychrotrophic bacteria belong to several genera that can multiply at low temperatures, but their optimum growth temperature can vary. Many are mesophilic and grow more slowly at lower temperatures. Due to the bulk storage of raw materials in industries, coconuts are sometimes processed only a few days after being stored. The storage period under refrigeration leads to an increase in the number of psychrotrophic microorganisms. High numbers will cause future defects in the storage of coconut water (BRITO, 2001).

High counts of psychrotrophic microorganisms are associated with deficiencies in hygiene, failures in cleaning and sanitizing and refrigeration equipment or when the refrigerated storage time is too long. When hygienic processing conditions are good, the count of psychrotrophic bacteria is low, but if conditions are poor they can account for 75% or more of the total bacterial population. Psychrotrophs have become more important due to the introduction of cold preservation technologies in the food chain (BRITO, 2001). According to the EFSA report (EFSA, 2010) on reported foodborne outbreaks in 2008, the main cause of foodborne infectious diseases was cross-contamination due to incorrect food handling (33.8%). The other causes identified as responsible for the occurrence of foodborne infectious diseases were: inadequate heat treatment (15.8%), use of contaminated unprocessed ingredients (9.2%) and disregard for storage time or temperatures

(5.6%).

According to Brasil (2009), the microbiological characteristics allowed for coconut water are as follows: Sum of molds and yeasts with a maximum of 20UFC / mL, Escherichia coli or thermotolerant coliforms of 1 UFC / mL, Salmonella sp with a minimum of absence in 25ml, the current legislation does not include analyzes for psychrotrophic bacteria.

2.1.13 Sensory evaluation

Meilgaard et al. (1987), reports that sensory analysis is a powerful tool for measuring and interpreting the reactions produced by food characteristics and the way they are perceived by the human sense organs. It consists of evoking, measuring, analyzing and interpreting reactions to food characteristics that are perceived by the sense of sight, smell, taste and hearing. Sensory methods are those that make it possible to evaluate the individual's impression of the condition of the product and its quality (DUTCOSKY, 2007).

Nogueira et al. (2004), reports that sensory analysis is an indispensable tool for assessing the quality of coconut water, because through it the consumer comes into direct contact with the product and can then determine its degree of acceptance or rejection. Aroucha et al. (2000) point out that this analysis is extremely important for assessing the quality of coconut water from different varieties and at any stage of fruit ripeness.

According to the F.I.P. (2004), sensory evaluation is a very important tool that can help industries to develop and control their products so that they are more acceptable to the end consumer. According to Brasil (2009), as far as sensory characteristics are concerned, coconut water, with the exception of concentrated and dehydrated coconut water, should have a characteristic color, a slightly sweet taste, its own aroma and an appearance varying from translucent to opaque. The presence of small amounts of supernatant particles from the coconut pulp does not disqualify the product.

2.1.14 Good Manufacturing Practice

According to Hoffman et al. (2002), coconut water is considered a sterile drink, but contact with the environment, utensils and equipment and handlers without proper sanitization can make it unsuitable for consumption, reducing its quality and even becoming a source of contamination by pathogenic microorganisms.

In recent decades, the health authorities of several countries have begun to introduce significant changes to their legislation, adopting and implementing new control instruments, through the principles of Good Manufacturing Practices and the Hazard Analysis and Critical Control Points (HACCP) system, based mainly on the recommendations of the *Codex Alimentarius* Commission.

These procedures focus on process control and food risk analysis, establishing critical processing limits and verifying compliance with them. For this to happen, the producer will need to have systems in place to record the industrial process and be able to demonstrate this through evidence, facts or proof that their products do not present significant risks to consumers (PAHO, 2003).

Since the food handler can be a vehicle for contamination, and the asymptomatic carrier is difficult to detect, it is essential to implement and maintain the HACCP system and Good Manufacturing Practices, which are the best ways of preventing Foodborne Diseases (GERMANO et al., 2008).

Hygiene conditions must be a constant concern in order to minimize contamination with microorganisms that could spoil the product. Attention must be paid to the personal hygiene and health of employees, as well as the cleanliness and maintenance of equipment and the working environment. The processing room and all equipment and utensils must be washed and sanitized daily before and after use (EMBRAPA, 2005). B P F is a program used to control processes and operating procedures with the aim of facilitating the operation of innocuous food, and covers procedures related to the use of facilities, reception and storage, maintenance of equipment, training and hygiene of workers, cleaning and disinfection, pest control and return of products (CRUZ et al., 2006).

GMP quality control is relevant to the public, industry and the government, as it is a program that checks that industrial processes and controls carried out in establishments are being implemented in such a way as to minimize and avoid risks to public health, preventing economic fraud and quality losses. For the public, this measure aims to ensure food safety by providing food that is safe in terms of microbiological quality; for the industry, it aims to control the quality of the products offered to the public by observing aspects such as taste, texture and appearance, as well as food safety; *for the* government, good manufacturing practices aim to establish essential hygiene requirements and good preparation practices (DUREK, 2005).

For the National Health Surveillance Agency (ANVISA), Good Manufacturing Practices are understood to be a set of measures that must be adopted by food industries in order to guarantee the sanitary quality and conformity of food products in relation to technical regulations (ANVISA, 2013).

As the Good Practices program almost always requires structural and behavioral changes, the commitment of management to the resources needed for implementation is essential, and the success or failure of this program will depend on the commitment made by management, which must be aware of the benefits and difficulties arising from the implementation of the program (LOVATTI, 2004).

The adoption of Good Manufacturing Practices (GMP) is the most feasible way to achieve adequate levels of food safety, helping to guarantee the quality of the end product. In addition to reducing risks, they enable a more efficient and satisfactory working environment, optimizing the production process, controlling possible sources of cross-contamination and guaranteeing product identity and quality specifications.

A GMP program covers a wide range of industrial aspects, from the quality of raw materials and ingredients, including product specifications, the selection of suppliers and water quality, as well as the recording of all agro-industry procedures on appropriate forms, to building and hygiene recommendations, adapted to the reality of the establishment.

According to Brasil (2009), Draft Normative Instruction No. 27, of July 2009, which governs the Identity and Quality Standards for coconut water, in Chapter III "Operational hygiene procedures", Section I Management of Good Manufacturing Practices, Art. 27; The product must be processed, packaged, stored, preserved and transported as determined by the management of Good Manufacturing Practices, in compliance with the specific legislation. And that the Good Manufacturing Practice management plan must be drawn up by the production manager. He or she must be responsible for approving it with the Ministry of Agriculture, Livestock and Food Supply; for managing it, controlling the implementation of the recommendations contained therein, as well as for the mandatory records on spreadsheets. The Good Manufacturing Practices management plan must complement the principles required by the specific legislation in force, paying special attention to the following items.

With regard to operational procedures, Article 28 states that the following procedures must be adopted by the coconut water handling establishment in addition to the other legal requirements:

The fruit should be examined for integrity and health and discarded if they show damage, stains or punctures caused by insects or tools. The fruit should be washed with drinking water under good pressure to remove dirt from the crop. They should then be sanitized in accordance with Good Manufacturing Practices. Once they have been selected and sanitized, they should be stored in a hygienically and sanitarily controlled environment, kept permanently clean, dry and ventilated, in order to avoid cross-contamination, in accordance with Good Manufacturing Practices.

Art. 29. As for the floors and drainage system, the coconut water handling establishment must comply with the following characteristics in addition to the other legal requirements:

Floors must be resistant, waterproof, non-absorbent, washable, slip-resistant and free of cracks and gaps. They must be permanently clean and free of fruit debris and other residues. They must have a sufficient slope for perfect drainage of wash water. The wash water drainage system, gutters and drains must be easy to clean and kept clean at all times.

Art. 30. With regard to equipment and utensils, the coconut water handling establishment must comply with the following characteristics in addition to the other legal requirements:

The equipment and tanks used to store and process the coconut water fruit must be made of material that does not endanger the safety of the raw material or the product being processed.

All equipment and utensils used in processing must be cleaned and disinfected before production begins, in accordance with Good Manufacturing Practice. The equipment used to pierce or cut the fruit must be made of material that does not endanger the safety of the product to be extracted and must be constructed in such a way as to allow it to be perfectly and constantly aseptic. Fixed equipment must be installed in such a way as to allow easy access for thorough cleaning.

Art. 31. When bottling coconut water, all stages must be carried out without delay and under hygienic and sanitary conditions that prevent the possibility of contamination, deterioration or the development of pathogenic microorganisms.

You should decide on an agro-industrial project that allows for a continuous flow of production, so that there is no contact between the processed product and the raw material in the processing environment.

The adoption of Good Manufacturing Practices (GMP) is one of the most important tools for achieving adequate levels of food safety and, with this, contributes significantly to guaranteeing the quality of the final product (EGAN et al., 2007).

2.1.15 Experimental planning

The lack of experimental planning is often the cause of research failure. Through experimental planning based on statistical principles, the maximum amount of useful information can be extracted from the system under study, using the minimum number of experiments (AZOUBEL, 2003).

The essence of good planning is to design an experiment in such a way that it is able to provide exactly the kind of information you are looking for. To do this, it is necessary to clearly define what objective the experiments are intended to achieve, because this will determine what type of experimental design should be used (BARROS NETO et al., 1995).

The Response Surface Methodology (RSM) consists of two distinct stages: modeling and displacement. These steps are repeated as many times as necessary, with the aim of achieving an optimum region (maximum or minimum) of the surface under investigation. Modeling is usually done by fitting linear or quadratic models to the results obtained in the experimental design. The displacement always occurs along the path of maximum ascent of a given model, which is the path on which the response varies most pronouncedly. The main attraction of this methodology is the reduction in the number of trials needed to assess the influence of factors on process responses (RODRIGUES, 2005).

3. MATERIAL AND METHODS

This work was carried out in the Physico-Chemical and Microbiological Food Analysis Laboratories of the Technological Vocational Center (CVT) and the Agro-Industrial Science and Technology Center of the Federal University of Campina Grande (UFCG), Campus Pombal - PB.

3.1 Raw materials

The raw material used was ana verde coconut water produced industrially and sold in the states of Paraiba and Cearà. Each plot or sample consisted of four 300 ml transparent PET cups. The raw material acquired by the industries does not undergo any post-harvest procedures.

3.2 Production of green dwarf coconut water bottled by industries

Figure 3.1 and Figure 3.2 show the flowchart of the production process for refrigerated, bottled green coconut water and the raw material, the dwarf green coconut used by the industries.

Figure 3.1 - Stages in the production process of refrigerated green coconut water bottled by the industries evaluated.

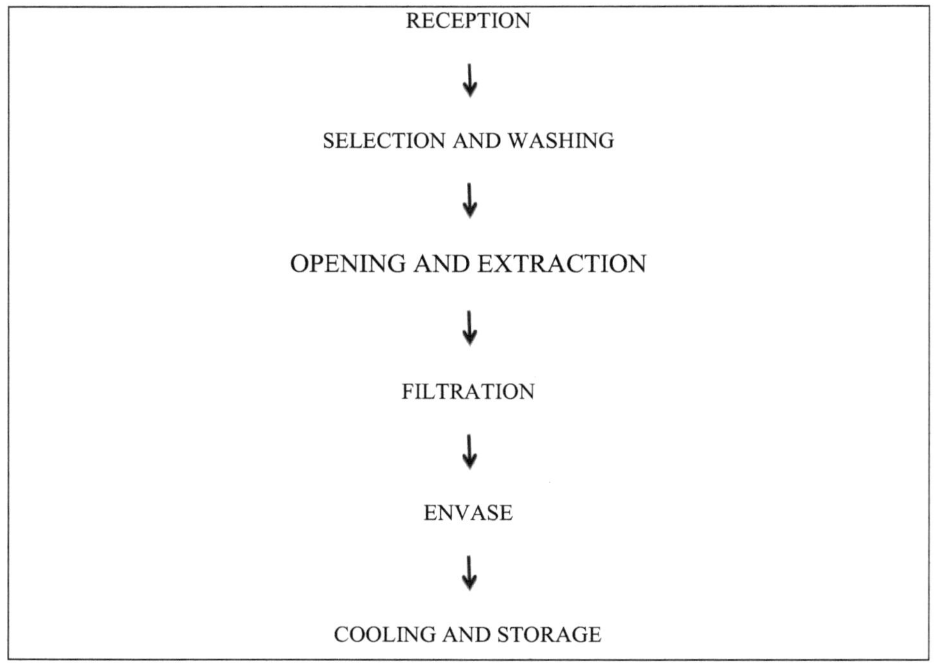

Figure 3.2- Green dwarf coconut used by industries.

Source: author

In the industries evaluated (coconut water), the coconuts were washed in a water tank with a capacity of 5,000 liters and tiled tanks with a capacity of 10,000 liters with a pressure jet of water to remove the dirt found in the raw material. They are then taken to the processing room, which is sanitized with 100 ppm hypochlorite. In the processing room, there is no air conditioning, making it a hot environment, conducive to the proliferation of microorganisms. The coconuts were cut using a stainless steel knife, and before opening the coconut, the physical condition of the coconut was analyzed.

The coconut water was collected in a transparent plastic bucket with a capacity of five liters and strained through a cotton cloth bag by the industry (A), filtering it twice before filling it manually. In industry (B), the coconut water was filled using a stainless steel dosing machine with three nozzles, adjusted to fill the container in previously defined quantities in which the coconut water has no contact with air. The coconut water is filtered through a fine mesh stainless steel sieve with an opening of 0.3 mm. The opening and extraction systems are mostly developed by the fillers themselves to meet their specific needs.

The cups were sealed with a hot-air sealing machine (lid sealing machine) by the industries.

The industry cups (A), after being sealed, were taken to a vertical freezer where they were stored at a temperature ranging from 5 °C to 8 °C, as shown in Figure 3.3. The industry cups (B), after being sealed, were taken to a refrigeration and freezing chamber with a temperature of between 4 °C and -12 °C, as shown in Figure 3.4. This operation should be carried out as quickly as possible to preserve the original characteristics of the water and prevent microbiological growth or changes in color. In the industrial processing room of industry (A), there are four employees to perform this function, spending an average of 10 to 15 minutes to fill 30 glasses. In the case of industry (B), the filling process is faster and they can fill 40 glasses in 15 minutes. It has five employees to process the filling and sealing of the cups. None of the industries cool the coconut water in a stainless steel tank to minimize the risk of contamination.

Figure 3.3 - Storage of coconut water processed by industry A in a vertical freezer.

Source: author

Figure: 3.4 - Storage of coconut water processed by industry B in a refrigeration chamber.

Source: author

The factories commercially transport the raw material in Styrofoam boxes with ice to local businesses and surrounding towns. The factories that manufacture coconut water in the interior of the Northeast are staffed by untrained employees with no technical manager capable of practicing

Good Manufacturing Practices. According to Cabral et al. (2005), green coconut water should be kept refrigerated until it is consumed. The recommended temperature for storage in cold rooms or refrigerators ranges from 6 °C to 10 °C. The shelf life of coconut water refrigerated at 6 °C is around 3 days, depending on the hygiene conditions used in processing, the time elapsed between opening the coconut and filling it, and the degree of ripeness of the fruit used as raw material. It is important that the cold chain is not broken during the entire distribution and sale of coconut water, until it is consumed, in order to guarantee the quality of the product.

3.3 Sample collection

Due to a lack of marketing registrations by the Ministry of Agriculture and Livestock, the states of Cearà and Paraiba, in the Sertao, only have two industrial units selling refrigerated coconut water, legalized by the inspection bodies.

The samples were collected from two industrial units on the day of manufacture, duly identified and transported to the laboratory in isothermal boxes with ice, where they were analyzed for physical, physico-chemical, microbiological and sensory parameters in order to assess the quality of coconut water immediately after processing.

3.4 Product storage

The samples were obtained from the industrial units and stored at temperatures of 7 °C, 12 °C and 2 °C, with 7 °C being the maximum temperature required by current legislation (BRASIL, 2009), 12 °C based on storage at points of sale, and 2 °C, the temperature under study.

The samples obtained from the two industries were stored according to the experimental factorial design $2^2 + 3$ repetitions at the center point, as shown in **Tables 1 and 2**, in order to quantitatively evaluate the influence of the input variables on the responses.

Table 3.1 - Real and coded values of the input variables for the storage of processed coconut water

Variables .	Levels		
	-1	0	1
X_1 - Temperature (°C)	2	7	12
X_2 - Time (day)	1	15	30

Table 3.2 - Matrix of the 2 factorial design^2 + 3 central points for the storage of processed coconut water

Experiments	Variables

	Temperature (°C)	Time (day)
	X_1	X_2
1	(-1) 2	(-1) 1
2	(+1)12	(-1) 1
3	(-1) 2	(+1) 30
4	(+1) 12	(+1) 30
5	(0) 7	(0) 15
6	(0) 7	(0) 15
7	(0) 7	(0) 15

The coconut water samples in 300 mL glasses corresponding to the experiments were stored in a BOD (incubator) with digital temperature control and the physical condition of the glasses was monitored every day of storage, as well as the temperature, in order to avoid possible storage failures.

The coconut water obtained after industrial processing was characterized chemically, physically, microbiologically and evaluated sensorially.

3.5 Physical and physico-chemical analysis of stored coconut water

3.5.1. Physical analysis

3.5.1.1 Turbidity (NTU)

Turbidity was determined by spectrophotometry at 610 nm (relative to distilled water). Policontrol AP 2000 turbidimeter. At this wavelength, the absorbance (A) was read for each sample. The transmittance values were calculated according to the ratio: $Tr = (10^{-A}) \times 100$ (CAMPOS et al., 1996).

Turbidity will be calculated from:

$T = 100 - Tr$

Where T is the transmission factor at a wavelength of 610 n

3.5.1.2 Conductivity $(mS/cm)^{-1}$

Determined by the electrometric method, which is based on the determination using the LUCADEMA conductivimeter, model Mca 150.

Conductivity analysis is based on contact between the electrode and the sample. The electrode is submerged in distilled water. For the experiment, a Becker was used with the samples.

3.5.1.3 Viscosity $(mm^2 /s)^{-1}$

A DV-E (Viscometer Brookfield) viscometer was used to determine the viscosity of the coconut water samples. The instrument is equipped with cylinders of different diameters (*spindles*), in which the appropriate cylinder is used according to the viscosity of the fluid. For the coconut water under analysis, the cylinder with an external diameter of 100 mm (reference *splindle* S2) was used.

3.5.1.4 Hydrogen Potential (pH)

Determined by the potentiometric method, which is based on determining the hydrogen concentration using the pH meter LUCADEMA, model mPA-210, followed by method 017/IV of the Adolfo Lutz Institute (2008), recommended by Brasil (2003).

3.5.1.5 Total Soluble Solids (°Brix)

The total soluble solids content expressed in °Brix was determined by the refractometric method, using a REICHERTAR 200 refractometer. With temperature correction, about 1 ml is added under the reader and then the reading is taken according to the analytical standards of the Adolfo Lutz Institute (BRASIL, 2008). The aim of this method is to assess the amount of Total Soluble Solids (TSS) using the °Brix scale (total measure of soluble solids in the sample to be analyzed). Soluble solids are basically sugars (sucrose, fructose and glucose) and so °Brix is basically considered to be the percentage of sugar present in the sample. "To refer to Brix we use the term 'degrees °Brix', which is equivalent to a percentage.

3.5.1.6 Total Titratable Acidity (%)

Titratable acidity (TA) was determined in duplicate using an aliquot of coconut water to which 50 mL of distilled water and 3 drops of 1% alcoholic phenolphthalein were added. The sample was then titrated with a previously standardized 0.1 N NaOH solution, and the results were expressed as a percentage (%) of citric acid. Total titratable acidity was determined using the acidimetric method in the Adolfo Lutz Institute manual (BRASIL, 2008).

3.5.1.7 Ascorbic Acid (mg/100g)

The ascorbic acid content was analyzed according to the AOAC (1997) method, modified by BENASSI; ANTUNES (1988), which uses 2g of the sample and 50 ml of 1% oxalic acid solution as the extracting solution, based on titration of the sample using 2,6 dichlorophenol-indophenol sodium, which gives a blue color in alkaline solutions and a pink color in acidic solutions. The results were expressed in mg of ascorbic acid/100 g of the sample.

3.5.1.8   Ash (%)

Ash was determined according to the methodology of the Adolfo Lutz Institute (BRASIL, 2008), using the gravimetric method by incineration in a muffle furnace at 500 °C, and the results were expressed as a percentage (%).

5g of the sample was placed in a porcelain capsule, previously heated in a QUIMIS muffle furnace (Q- 318S24) to 550°C, cooled in a desiccator to room temperature and weighed. The sample is carbonized on a hot plate, then incinerated in a muffle furnace at 200 °C, increasing the temperature until the charcoal is completely removed. The sample is cooled to room temperature in a desiccator and the porcelain capsules are weighed.

3.5.1.9   Protein (%)

The protein content was determined using the method described by the Adolfo Lutz Institute (BRASIL, 2008) using a factor of 6.25.

The protein content was determined by the Kjeldahl method, using acid digestion, followed by distillation and titration. The method is based on heating the sample with sulphuric acid for digestion until the carbon and hydrogen are oxidized. The nitrogen in the protein is reduced and transformed into ammonium sulphate. After digestion, 80 ml of water, 5 drops of phenolphthalein, 0.5 g of zinc powder and 80 ml of 40% NaOH were added, then the flask was connected to the distiller with its end immersed in an erlenmeyer flask containing 50 ml of 0.1N H2SO4 and 3 drops of methyl red indicator, where about 100 ml was distilled. After distillation, titration was carried out with a 0.1 N NaOH solution until the indicator turned yellow.

3.5.1.10   Reducing sugars (%)

The reducing sugar content was determined using the alkaline reduction method described by the Adolfo Lutz Institute (BRASIL, 2008), and the results expressed as a percentage of glucose. The sample was weighed into a beaker in which 50 ml of distilled water was added, then placed in a water bath at 80 °C for 15 minutes, where the sugars were extracted. After this time, the sample was transferred to a 100 ml volumetric flask for calibration, then filtered through qualitative filter paper into a 250 ml beaker. This filtrate was used to titrate the fheling A and B solutions in an erlenmeyer flask containing 10 ml of each solution, 50 ml of water and 3 drops of methylene blue indicator. The solution contained in the flask was heated on a hotplate while stirring and titrated with the filtrate until the indicator turned brick red. All determinations were made in triplicate.

3.6 Microbiological analysis during product storage

3.6.1.. Thermotolerant coliforms

The multiple tube technique (Most Probable Number - MPN) was used to determine thermotolerant coliforms, following the methodology (BRASIL, 2003).

The Most Probable Number (MPN) technique was used, using a series of 3 tubes, making successive dilutions ($10^{\wedge 1}$, 10^{-2}, 10^{-3}). Aliquots of 1 mL were transferred to tubes containing Bile Brilliant Green Broth and incubated at 35 °C ± 1°/24 hours for the determination of coliforms at 35 °C. Positive tubes were those showing turbidity and gas production. Quantification of coliforms at 45°C consisted of transferring aliquots of all the positive tubes from the Coliforms at 35°C culture to tubes containing Escherichiacoli (EC) *broth* and incubating them at 45°C /±1°C for 48 hours in a water bath with constant agitation and temperature to confirm Coliforms at 45°C.

3.6.2. Salmonella

Salmonella sp. was tested using the methodology indicated in (BRASIL, 2003).

To determine *Salmonella* sp, 25 g (mL) of each sample was weighed and homogenized in 225 mL of 0.1% peptone water to obtain a dilution of $10^{\wedge 1}$. From this dilution, 0.1 mL aliquots were taken and successive dilutions (10^{-2}, 10^{-3}) of each dilution were inoculated and spread on the surface of Rambarch Agar using a *Drigalski loop*. The plates were incubated for 48 hours at 37 °C.

3.6.3. Molds and yeasts

They were based on the methodology indicated in (BRASIL, 2003). Standard procedure: they were carried out in duplicate for each dilution, 10^{-1}, 10^{-2}, 10^{-3}. Thus, 1 mL of each dilution was transferred to sterilized Petri dishes, duly identified. Next, 10 mL of nutrient agar, previously melted and cooled to 50 °C, was added to each plate, and the plates were gently homogenized with circular movements in the shape of an "8" for 8 times. After solidifying at room temperature, the plates were inverted and incubated at 35 °C ± 1 °C for 48 hours.

3.6.4. Psychrotrophic

They were based on the methodology indicated in (BRASIL, 2003). Standard procedure: they were carried out in duplicate for each dilution, which was 10^{-1}, $10{-}2$, $10{-}3$. Thus, 1 mL of each dilution was transferred to sterilized Petri dishes, duly identified. Then, 10 mL of nutrient agar, previously melted and cooled to 50 °C, was added to each plate, and the plates were gently homogenized with circular movements in the shape of an "8" for 8 times.

After the nutrient agar had solidified at room temperature, the Petri dishes were incubated inverted at 7 °C ± 1 °C /8 days.

3.7 Sensory analysis of processed coconut water

The coconut water samples from the factories were sensorially analyzed at time zero, using a comparison test with a 9-point hedonic scale (9 = extremely liked and 1 disliked), according to the Adolfo Lutz Institute (IAL, 2008) with some adaptations. The attributes evaluated were product appearance, product aroma, product flavor, attitude towards buying the product and product preference. The form used for the sensory analysis of the samples can be found in Appendix A.

The sensory evaluation was carried out in the microbiology laboratory of the CVT in the city of Pombal. It was planned so that each of the participants tasted the 4 samples served sequentially, balanced in relation to the order of presentation.

A purchase intention test was also applied, in accordance with the methodology described by the Adolfo Lutz Institute (IAL, 2008), which states that by means of scales or purchase intention, the individual expresses their desire to consume, acquire or buy a product offered to them using a structured 5-point scale (1 = certainly would buy; 2 = Possibly wouldn't buy; 3 = Maybe would buy / maybe wouldn't buy; 4 = Probably would buy; 5 = Certainly would buy). And in terms of purchase preference (first, second, third, fourth).

The samples were served in disposable transparent plastic cups under ambient lighting and coded with a 3-digit number, chosen at random so that the tasters would not be influenced by the external environment. The samples were presented in a balanced way so that they appeared in each position an equal number of times. Forty untrained male and female tasters aged between 15 and 20 years were used, representing a public school in the city of Pombal (P.B), where they were given a brief explanation of how to proceed with their evaluations. The data obtained from the acceptance test was statistically analyzed using the Tukey test at 5% significance level.

3.8 Statistical analysis

The results obtained from the experimental factorial design were statistically analyzed using ANOVA (analysis of variance) and the response surface method, using the statistical program STATISTICA® version 5.0 (2004). ASSISTAT version 7.5 Beta was also used. Silva; Azevedo, (2009), to check for possible statistical differences between the parameters determined in coconut water. Tukey's test at 5% probability was used to compare the means.

4. RESULTS AND DISCUSSIONS

This section contains the results and discussion of the physical, physico-chemical, microbiological and sensory characterization of ana verde coconut water filled and sold by industries in the Paraiba and Cearà hinterlands.

4.1 . Characterization of the raw material

Tables 4.1.1 and 4.1.2 show the average values of the physical and physical-chemical parameters used to characterize the ana verde coconut water produced and marketed by factories A and B.

Table 4.1.1 Results of the physico-chemical characterization of ana verde coconut water produced commercially by the industries.

Exp.	Turbidity (N.T. U)	Conductivity (mS/cm)$^{-1}$	Viscosity (mm^2/s)$^{-1}$	pH	SST (°Brix)	ATT (g/100 ml)
Ind. A	9,50b	4,38 a	0,32a	5,05a	5,35a	0,07a
Ind. B	18,1a	4,39 a	0,30a	5,15a	5,45a	0,03a
DMS	6,63293	3,55998	0,48052	0,30450	1,85220	0,13306
MG	13,8	4,38	0,85	5,10	5,40	0,05
CV%	11,14	18,84	14,22	1,39	7,97	54,7

MSD - Minimum significant deviation; MG - General average; CV - Coefficient of variation.

Averages followed by different letters, lower case in the columns and upper case in the rows, differ according to the Tukey test (p ≤ 0.05).

Table 4.1.1 shows the average turbidity values of the coconut water analyzed. Industry A had a value of (9.5 N.T.U) and industry B had a value of (18.1 N.T.U). These values were close to the values of 12.1 and 13 (N.T. U) reported by Silva et al (2009), and 7.8 and 13.8 (N.T.U) and Magalhaes et al. (2005). The aforementioned turbidity values showed statistically significant differences according to the Tuckey test applied at a 5% probability level. This variation in the turbidity present in the coconut water samples analyzed is probably due to the fact that they have not undergone any post-harvest processes, such as inadequate handling and improper storage conditions for the raw material, which accelerate the deterioration process of coconut water, reducing its useful life.

In relation to the average conductivity values of this study, industry A presented a value of (4.38 mS/cm-1) and industry B a value of (4.39 mS/cm^{-1}), similar values were found by Silva et al. (2010), who studied the characterization of fresh and industrialized coconut water sold in the Paraíba sertao, and found results ranging from (4.18 mS/cm^{-1} to 5.47 mS/cm^{-1}). It can be seen that the conductivity values did not show significant differences according to the Tukey test at a 5% probability level, thus expressing the amount of dissolved salts in which these salts present in coconut water may vary according to the climate, type of soil and cultural treatments during the fruit's development period.

With regard to the viscosity of coconut water, a small variation was found between (0.32 mm^2/s^{-1}) for industry A and (0.30 mm^2/s^{-1}) for industry B. Higher values were found by Pinheiro et al.(2009), when they studied different brands of coconut water obtained by the aseptic process, which showed a small variation between (1.02 mm^2/s^{-1} and 1.08 mm^2/s^{-1}). The viscosity of many liquids changes with changes in temperature; high temperatures generally decrease viscosity because the friction between molecules decreases; coconut water from industry A has a higher viscosity than from industry B, there may have been a temperature oscillation during storage.

Comparing the average pH values of the coconut water in this study, which ranged from (5.05) for industry A to (5.15) for industry B, with the results obtained in research carried out with other coconut waters, lower pH results were found by Silva et al. (2010), ranging from 4.81 to 4.82. Rosa and Abreu (2002) found a value of 4.91 for 7-month-old fruit of the green dwarf variety. The values in this study are not in line with current legislation (Brasil, 2009), with the permitted range being 4.3 to 4.5, which is above the permitted value. According to Embrapa Semi- Arido, when carrying out experiments with water from seven-month-old green dwarf coconuts, the pH was around 5.0 and the sugar and mineral salt content was more balanced, giving them a more pleasant taste due to a reduction in astringency. In six-month-old fruit, the pH value of coconut water was 4.5, making it more acidic and astringent. According to Nery et al. (2002), evaluating the pH is important because the desired sweet taste and astringency are achieved with a pH close to 5.5. These values were observed in fruit from six dwarf coconut cultivars around seven months old. The pH of coconut water varies little over the course of the fruit's development, changing with ripening and increasing as it progresses. The pH range found in the literature for coconut water varies from 4.5 to 5.7

regardless of the varieties (LIRA, 2010). The pH of the present study is close to the pH values of mature coconut water, around 5.1 to 5.2. This pH value may also be due to the fact that the raw material does not undergo post-harvest procedures, such as the non-uniform stage of ripeness of the fruit during harvesting for its industrialization.

The soluble solids content of the coconut water for factories A and B ranged from 5.35 to 5.45 °Brix, respectively, and is therefore within the standards set by the legislation in force in Brazil (2009), where soluble solids can vary from 4.5 to 7.0 °Brix. Aroucha et al. (2005), found similar results to this research for the total soluble solids content depending on the stage of ripeness of the ana verde and ana vermelho varieties, in the 8th month showing values of 5.4 °Brix for both varieties. Rosa; Abreu (2000), reports that in green coconuts, one of the parameters for harvesting fruit for consumption of coconut water should be around 6 °Brix, a value usually found in fruit 6 to 7 months old, at which time coconut water has a higher amount of non-reducing sugars (fructose). For Lira (2010), the most palatable °Brix and pH in his study on commercial sterilization of green coconut water in the region of Sousa-PB, is between the 6th month of age of the fruit after natural inflorescence, when there is a higher water content in the fruit, an increase in °Brix and a reduction in pH, indicating the ideal period for consumption of green coconut water, resulting in a light and sweet taste.

The average acidity content of the coconut water samples ranged from (0.07g/100 ml) for industry A to (0.03g/100 ml) for industry B. Silva et al. (2010) found higher results for the acidity content, ranging from 0.11 g/100 ml to 0.012 g/100 ml. The legislation in force in Brazil (2009), for the acidity of green dwarf coconut water, has maximum and minimum limits of (0.18 g/100 ml and 0.06g/100 ml) respectively, with industry B therefore being below the permitted value, outside the standards established by the legislation in force. According to Stumbo (1973), foods with a pH higher than 4.5 are considered low acid foods. Srebernich (1998) states that the main acid present in coconut water is malic acid. Other acids are present, but they represent very little in the composition of coconut water.

Table 4.1.2 Results of the physico-chemical characterization of commercially produced dwarf green coconut water.

Exp.	Vit. C Mg/100g)	Ash (%)	Protein (%)	Sugars Reducers
Ind. A	1,60a	0,48[a]	0,20a	2,25a
Ind. B	1,25a	0,47[a]	0,20a	2,50a
DMS	2,79908	0,03900	0,43063	0,76720
MG	1,42	0,48	0,20000	2,38
CV%	45,61	1,88	0,50	7,48

MSD - Minimum significant deviation; MG - General average; CV - Coefficient of variation.

Averages followed by different letters, lower case in the columns and upper case in the rows, differ according to the Tukey test (p ≤ 0.05).

Table 4.1.2 shows the physical and chemical values for the characterization of coconut water filled and marketed by the industries. It can be seen that there were no significant differences in any of the coconut water samples analyzed using the tuckey test at a 5% probability level.

51

As for the vitamin C content found in coconut water, it varied between (1.60 mg/ 100 g) for industry A and (1.25 mg/ 100 g) for industry B. Lower values were detected by Pinheiro et al. (2005), who studied the chemical, physico-chemical, microbiological and sensory characterization of different brands of coconut water obtained through the aseptic process, and reported values ranging from (0.17 mg/ 100 g to 0.23 mg/ 100ml) respectively. The possible cause of the low concentration of this parameter may be due to oxidation of the coconut water during the collection of the water from the fruit or during processing, given that Rosa; Abreu (2000), found similar values to this study, (1.20 mg/ 100 ml), Tavares et al. (1988), states that coconut water is rich in minerals and vitamin C in six-month-old fruit, as is the case with the yellow dwarf and red dwarf cultivars of Malassia, which are considered a source of vitamin C. According to Brasil (2009), the addition of citric acid to correct acidity is permitted in the case of coconut water that has undergone some technological preservation process, in accordance with specific legislation for essential nutrients.

The ash content showed values of 0.48% for industry A and 0.47% for industry B. These values were close to those reported by Lira, (2010), who, when studying the commercial sterilization process of green coconut water using ceramic membranes, found values close to those of this study, ranging from 0.34 to 0.40%, showing that this is a food rich in mineral salts. Although an analysis of the mineral constituents contained in coconut water was not carried out. Rosa; Abreu (2000), in their studies, found reasonable amounts of phosphorus (7.4 mg/ 100 g), sodium (7.05 mg/ 100 g), potassium (156.86 mg/100 g) and also magnesium, manganese and iron.

The protein contents found in the green coconut water samples analyzed in this study were (0.20 mg / 100g), respectively for the industries, lower values were found by Pinheiro et al. (2005), studying the chemical, physical-chemical, microbiological and sensory characterization of different brands of coconut water obtained by the aseptic process (0.07% to 0.02% and 0.11%). According to Rosa; Abreu (2000), the protein content of natural coconut water is around 370 mg/100mL. Coconut water is low in protein, with a higher concentration at maturity in the 11th to 12th month when the fat content is more present. Lower values were found by Lira (2010) in his study, ranging from 0.06 to 0.08%. This low protein content indicates that coconut water is low in protein, so it is recommended as an aid in the diet of recovering patients who need to absorb mineral salts and control their protein intake.

The reducing sugar content ranged from (2.25%) for industry A to (2.50%) for industry B. Higher results were found by Costa et.al. (2006), which ranged from (6.67%) to (4.62%), and values of (0.57%) to (3.53%) were found by Pinheiro et.al. (2005), which are close to those of coconut water in this study. According to Vasconcelos (2000), sugars are used as a respiratory substrate, but are found in fruits in much higher quantities than those needed to generate energy. In coconut water, studies show that there is only a 2% drop in sugar content over a period of 7 to 12 months. This is because when the fruit is green, the sucrose units are not combined and there are sufficient quantities of free fructose. As time goes by, glucose and fructose combine to form sucrose, favoring a drop in sugar content. According to Ariola et al. (1980), this composition can vary between cultivars and within the same cultivar, depending on climatic conditions, soil fertility, time of year and stage of maturity. For Rosa and Abreu (2000), from this period of 6 and 7 months onwards, the volume of coconut water begins to decrease, which is accompanied by a reduction in the content of non-reducing sugars, thus influencing palatability, and by an increase in reducing sugars (sucrose), making the consumption of coconut water unsuitable for human health.

The traces of non-reducing sugars show that the samples are at an early stage of ripeness, where fructose and glucose (reducing sugars) predominate and sucrose (non-reducing sugar) is almost absent.This shows that the fruit was harvested up to the 7th month, which is the ideal stage for marketing green coconut water, according to Silva (2003), as there is a greater amount of free fructose (a higher sweetness content than sucrose), making the water sweeter.

4.2 Storage in Industry A

Table 4.2 shows the average values of the storage responses, physical and chemical pH, turbidity, conductivity, total soluble solids (TSS), total titratable acidity (TTA), vitamin C and viscosity of the experiments with coconut water produced commercially by the industry.

Table 4.2- Average values of the storage and physical-chemical responses of the experiments with coconut water produced commercially by industry A

Exp.	pH	Turbidity (N.T. U)	Conductivity $(mS/cm)^{-1}$	SST (°Brix)	ATT (% citric acid)	Vit. C (mg/100g)	Viscosity mm^2/s^{-1}
1	5,8 a	8,6 d	4,1 a	5.4 ab	1,5 e	1,1 a	0,32 c
2	5,1 c	9,5 d	4,1 a	5.8 ab	1,1 e	1,2 a	0,30 c
3	5,4 b	114,0 b	3,8 b	6,1 a	3,3d	0,7 a	0.6 bc
4	4,1 e	248,5 a	3,3 c	4,2 c	11,1 a	0,6 a	0,81 b
5	4,7 d	27,6 c	3,4 c	5,3 b	5,8 c	0,6 a	1,37 a
6	4,7 d	24,6 c	3.5 bc	5.5 ab	7,1 b	0,7 a	1,50 a
7	4,6 d	25,3 c	4,1 a	5.5 ab	6.0 bc	0,7 a	1.05 ab
DMS	0,11	6,42	0,26	0,80	1,25	0,72	0,48

MG	4,92	65,45	3,80	5,43	5,15	0,81	0,85
CV(%)	0,54	2,50	1,72	3,71	6,12	22,2	14,2

MSD - Minimum significant deviation; MG - General average; CV - Coefficient of variation.

Averages followed by different letters, lower case in the columns and upper case in the rows, differ according to the Tukey test ($p \leq 0.05$).

The coconut water samples analyzed had average pH values ranging from 4.1 to 5.8. Experiment 1 outperformed the other experiments with a pH of 5.8. Experiments 5, 6 and 7 showed no significant differences, with values of 4.7, 4.7 and 4.6. Experiment 4 had the lowest value of 4.1, which is not in line with current legislation in Brazil (2009), with the permitted range of 4.3 to 4.5 being below the permitted value. But close to the values found by Rosa; Abreu (2000), in the range of 5.5 in their study with green dwarf coconut and by Tan et al. (2013), who found pH values for green coconut water of 4.78.

According to Lira (2010), in his study on the commercial sterilization process of green coconut water using ceramic membranes, he found similar pH results, above 4.7. This pH value is susceptible to the proliferation of pathogenic bacteria. However, it is outside the standards for chilled, pasteurized and frozen coconut water samples, which should have a pH between 4.3 and 4.5.Acidity and pH, as well as playing an important role in the germination of the fruit, vary considerably in the composition of green dwarf coconut water during the growth and ripening process, which means that coconut water has different qualities depending on this variation.

According to the data in Table 4.2, higher turbidity values were obtained in experiments 3 and 4, which had values of 114 and 248.5 N.T.U respectively, while experiments 1 and 2 had lower values for this variable, 8.6 and 9.5 N.T.U. Likewise, experiments 5, 6 and 7 did not show significant differences. According to Santos (2003), the turbidity of a sample is related to the transparency of the water, due to the presence of impurities that make it cloudy. The values obtained in this work in experiments 3, 4, 5, 6 and 7 were higher than those obtained by Silva et al. (2009), who, when studying coconut water from the fruits of the green dwarf coconut produced in the municipality of Trairi-CE, found levels of 13.80 and 12.08 N.T.U, while experiments 1 and 2 showed lower values. The presence of suspended solids during storage may have caused the higher values found in this study, changing the color of the water as well as the presence of bacteria and yeasts in suspension, resulting in a rapid increase in turbidity, which is why this parameter is an excellent indicator of the presence of these microorganisms in suspension.

In this study, the conductivity of the coconut water samples analyzed ranged from 3.3 to 4.1 mS/cm. Experiments 1, 2 and 7 showed no significant differences, as did experiments 4 and 5, and 6 and 3. It can be seen that there was not much change in the amount of salts present in the samples analyzed.

The values found in this work were lower than those obtained by Pinheiro et al. (2009), who observed electrical conductivity values between 5.09 and 8.13mS/cm when analyzing different brands of coconut water obtained through the aseptic process.

The total soluble solids content of the samples ranged from 4.2 to 6.1° Brix. Experiments 1, 2, 6 and 7 showed no significant differences and obtained averages of 5.4, 5.8, 5.5 and 5.5 respectively. The highest value for this variable was found in experiment 3 (6.1° Brix) and the lowest in experiment 4 (4.2° Brix). According to Brasil (2009), the soluble solids content of refrigerated green coconut water should be between 4.5 and 7.0 ° Brix, so all the experiments studied were within the established standards. With the exception of the total soluble solids in experiment 4, which obtained a lower value, it is possible that during storage there were changes in the composition of the coconut water, showing alterations in the sugar content present, with the total and free sugar content decreasing as the temperature increased. In experiment 3, we can see that there was a concentration of sugars in coconut water with temperature and storage time. The total soluble solids levels of the samples analyzed are lower than those found by Costa et al. (2006), who evaluated coconut water obtained by different preservation methods and found soluble solids ranging from 7.0 for chilled water to 7.30 for frozen water. Lira (2010), in his study, found soluble solids ranging from 6.0 to 6.7 °Brix. In a study carried out by Tan et al. (2013) with green coconut water, the soluble solids values found were close, at 5.60 °Brix, when compared to this work.

According to the ATT data in table 4.2, the highest content of this variable can be seen in experiment 4 (11.1%), while the lowest values were observed in experiments 1 and 2 (1.5 and 1.1%). In the other experiments 3, 5, 6 and 7, ATT ranged from (3.3%) in experiment 3 to (7.1%) in experiment 6.According to Brasil (2009), which establishes ATT values for chilled coconut water between 0.06g/100mL and 0.18 g/mL, the data from the experiments studied are above the quality standards for human consumption.

SREBERNICH (1998) states that the main acid present in coconut water is malic acid. Other acids are present, but account for very little of the composition of coconut water. Brazilian legislation allows the addition of citric acid for acidity in the case of coconut water that has undergone some technological preservation process (BRASIL, 2009). The content of fresh and processed green coconut water ranges from 0.18 to 0.31% according to (ROSA; ABREU 2000).

The vitamin C content ranged from (0.6 mg/ 100 g) to (1.2 mg/ 100 g). The highest values were obtained in experiment 2 and the lowest in experiments 3 and 4 (0.7 mg/ 100 g and 0.6 mg/ 100 g). According to Aragao et al. (2001), as the fruit of the green dwarf cultivar ripened, they observed an increase in the ascorbic acid content in the water as the fruit ripened, although the maximum value detected was (2.99 mg/ 100 g). This is higher than the value found in this study. The low level of vitamin C in the samples analyzed may have been due to processing. The highest concentration of vitamin C in green coconut water is in the sixth month of fruit ripening, according to VIANNA et al, (2008).

The viscosity of the coconut waters analyzed varied between 0.30 and 1.50 mm^2/s^{-1}. Experiments 5, 6 and 7 had higher viscosity values, while experiments 1, 2 and 3 had lower levels of this variable. PINHEIRO et. al. (2009), studying different brands of coconut water obtained in the aseptic process, found viscosity values between (1.02 and 1.08 mm^2/s)$.^{-1}$

Temperature is a parameter related to the internal energy of a substance. Several studies have shown that the viscosity of a liquid is highly influenced by changes in temperature, causing losses in product quality (OLIVEIRA; BARROS; ROSSI, 2009).

4. 3 Storage in Industry B

Table 4.3 shows the average values of the storage responses, physical and chemical pH, turbidity, conductivity, total soluble solids (TSS), total titratable acidity (TTA), vitamin C and viscosity of the experiments with coconut water produced commercially by the industries.

Table 4.3 Mean **values of** the physico-chemical storage responses from the experiments with coconut water processed by industry B.

Exp.	pH	Turbidity (N.T.U)	Conductivity (Ms/cma)	SST (°Brix)	ATT (% citric acid)	Vit. C (mg/100g)	Viscosity mm2/s-1
1	5.3 a	16.7 c	4.2 a	6.3 b	1.3 c	1.1 ab	0.34 c
2	5.2 a	18.1 c	4.2 a	6.2 b	1.2 c	1.2 a	0.34 c
3	5.2 a	68.2 b	3.8 b	6.8 a	1.7 c	0.65 ab	0.43 bc
4	3.8 d	271.0 a	3.8 b	4.3 d	11. a	0.65 ab	0.78 a
5	4.1 c	25.5 c	4.1 ab	5.7 c	7.7 b	0.60 b	0.50 b
6	4.1 c	25.4 c	4.1 ab	5.5 c	6.7 b	0.60 b	0.45 bc
7	4.6 b	26.5 c	4.1 ab	5.5 c	7.2 b	0.95 ab	0.47 bc
DMS	0.20	13.09	0.32	0.42	1.11	0.63	0.13
MG	4.64	64.5	4.06	5.78	5.41	0.82857	0.47
CV(%)	1.11	5.13	1.99	1.85	5.18	19.35	6.99

MSD - Minimum significant deviation; MG - General average; CV - Coefficient of variation.

Table 4.3 shows that the pH of the coconut water samples ranged from 3.8 to 5.3. Experiments 1, 2 and 3 were superior to the other treatments, with pH values of 5.3, 5.2 and 5.2, while experiment 4 had the lowest pH value (3.8). Experiments 5 and 6 had the same pH (4.1) and experiment 7 had a lower pH (4.6).

Experiments 1, 2, 3 and 7 are not in compliance with current legislation (BRASIL, 2009), with values higher than the legislation, Brazil (2009). Experiments 4, 5 and 6 are non-compliant, with values lower than the legislation, the permitted range being 4.3 to 4.5. Magalhaes et al. (2005) found pH values close to those of this study, in the range of 4.0 to 5.6. And close to the values found by Fernandes et al. (2011), who compared the physical and chemical characteristics of natural and industrialized green coconut water and obtained values ranging from 4.5 to 4.9. According to Kwiatkowski et al. (2012), analyzing the enzymatic activity and physical-chemical parameters of coconut water at different stages of development and climatic season, they observed that as the fruit's development time increases, the pH increases, varying with the age of the fruit, and can reach values of 4.7 to 4.8 in the first five months, and usually continues to result in values above 5.0 until the end of the fruit's development.

According to the data in table 4.3, the highest turbidity value was obtained in experiment 4 (271.0 N.T.U), while experiments 5, 6 and 7 had varying values (25.5, 25.4 and 26.5 N.T.U, respectively). Similarly, experiments 1 and 2 obtained lower values (16.7 and 18.1 N.T.U, respectively), while experiment 3 differed from the other experiments with a value of 6.8 N.T.U. The values obtained in this work in all the experiments were higher than those obtained by Silva et al. (2009), who, when studying coconut water from the fruits of the green dwarf coconut, produced in the municipality of Trairi-CE, found levels of (13.80 and 12.08 N.T.U). In a study carried out by Tan et al. (2013), where the turbidity of green coconut water was determined, values were also lower (4.85 to 5.60) than those found in this study. According to Resende (2007), when these suspended solids are composed of metal ions, and coconut water is exposed to the air, heavily manipulated (filtering, cooling), the metal ions react with phenolic compounds, or act as cofactors for enzymes, developing the coloration in the water. The pink color, which is the most common, can be attributed to the presence of high concentrations of magnesium ion. When there is an excess of the mineral iron in coconut water, the color develops to purplish, and in manganese it is black. Penha et al. (2005), in their studies on green dwarf coconuts, found that the opacity of coconut water is due to the presence of lipids in its composition, when the fruit is in a good state of preservation for consumption. According to Leber (2001), turbidity is the lack of transparency of a liquid due to the presence of suspended solids. The more solids in suspension, the cloudier the water and the higher the turbidity.

The conductivity values of the coconut water samples analyzed ranged from (3.8 to 4.2 mS/cm^{-1}). Experiments 1 and 2 obtained values of (4.2 mS/cm^{-1}). Experiments 5, 6 and 7 had conductivity equal to (4.1 mS/cm^{-1}), while experiments 3 and 4 had the lowest values (3.8 mS/cm^{-1}). The values found in this work were lower than those obtained by Pinheiro et al. (2009), who observed electrical conductivity values between 5.09 and 8.13 mS/cm-1 when analyzing different brands of coconut water obtained through the aseptic process. Fernandes et al. (2011), in their study of natural and industrialized coconut water, found data for natural coconut water ranging from 7.142 mS/cm^{-1} to 6.162 mS/cm^{-1} and industrialized water ranging from 6.812 mS/cm^{-1} to 5.672 mS/cm^{-1} .The high levels of conductivity are due to the ions present in coconut water, such as sodium, potassium, calcium and magnesium ions.

The total soluble solids content of the coconut waters analyzed ranged from 4.3 to 6.8 °Brix. Experiments 5, 6 and 7 obtained averages of 5.7, 5.5 and 5.5 °Brix. The highest value for this variable was found in experiment 3 (6.8 °Brix), and the lowest in experiment 4 (4.3 °Brix). According to Brasil (2009), the soluble solids content of coconut water should be between 4.5 and 7.0, so all the experiments studied are within the established standard, with the exception of experiment 4 which obtained a value lower than that established by the legislation, possibly, during processing, there may have been some change in the composition of coconut water, especially in the sugars present. This is in line with Tavares et al. (1998), who, when evaluating dwarf coconut water in the 12th month of maturation, found contents ranging from 3.1 to 7.1 °Brix. According to Fernandes et al. (2011), in their study of natural and industrialized coconut water, they found data for natural coconut water ranging from 5.79 to 4.22 and industrialized water ranging from 6.71 to 4.41 °Brix, values close to those found in this study, for whom the high levels of total soluble solids are probably due to the presence of sucrose.

According to the total titratable acidity data in table 4.3, the highest levels of this variable were observed in experiment 4 (11%), while the lowest values were observed in experiments 1, 2 and 3 (1.3, 1.2 and 1.7%, respectively). In the other experiments 5, 6 and 7, the ATT ranged from (7.7%) in experiment 5, (6.7%) in experiment 6 and (7.2%) in experiment 7. According to the legislation in force in Brazil (2009), which establishes total titratable acidity values for chilled coconut water of between 0.06 g/100 ml and 0.18 g/ml, the data from the experiments studied are above the quality standards for human consumption. Carvalho et al. (2006), in their study of the nutritional and functional properties and processing of coconut water during the 5th to 12th month stages of maturation of the green dwarf coconut, presented acidity values ranging from 0.5 to 1.5%, results which were also found by TAVARES et al. (1998).

Among the experiments analyzed, the vitamin C content varied from 0.60 to 1.2 mg/100 g. The highest value was obtained in experiment 2 (1.2 mg/100 g), and the lowest in experiments 5 and 6 (0.60 mg/100 g). Experiments 3, 4 and 7 showed no significant differences (0.65, 0.65 and 0.95 mg/100 g, respectively).According to Tavares et al. (1998), coconut water is rich in minerals, regardless of the age of the fruit, and in vitamin C. The largest varieties considered to be a source of vitamin C are the Anao Amarelo de Gramame and Anao Vermelho da Malàsia cultivars. The vitamin C values in this study are lower than those found by Aragao et al. (2001), which is 2.99 mg/100 g. However, the values obtained from the coconut waters analyzed in this work were higher than those found by PINHEIRO et al. (2009), who, when carrying out the chemical, physico-chemical, microbiological and sensory characterization of different brands of coconut water obtained by the aseptic process, found values ranging from 0.17 to 0.23 mg/100g. Rosa; Abreu (2000), found vitamin C values of 1.20 mg/100g in their studies, values close to those found in this work for experiments 1 and 2. Possible changes in the vitamin C content could be associated with possible protein denaturation or oxidation during the water collection operations.

The results for viscosity in Table 4.3 show a range from 0.34 to 0.78 mm^2 /s^{-1}). Experiments 3, 6 and 7 obtained values of (0.43, 0.45 and 0.47 mm^2 /s^{-1}). Experiments 1 and 2 had lower levels of this variable with a value of (0.34 mm^2 /s^{-1}). Experiment 4 had the highest value (0.78 mm^2 /s^{-1}) and experiment 5 (0.50 mm^2 /s^{-1}). Pinheiro et al. (2009), studying different brands of coconut water obtained in the aseptic process, found viscosity values between (1.02 and 1.08 mm^2 /s^{-1}), which were higher than those found in the present study.

4.4 Statistical analysis of coconut water stored for 30 days

In order to obtain a better statistical analysis of the results obtained in the storage of refrigerated coconut water, an evaluation of the variables time and temperature in the refrigerated storage condition was carried out through a factorial experimental design on the quality of processed and commercialized coconut water;

The results were analyzed using statistical methods, using the STATISTICA® version 5.0 (2004) statistical program, according to a 2 factorial design[3] with three repetitions at the central point, where the matrix with the input and response variables are shown in Table. 3.1 e 3.2

Table 4.4.1 shows the analysis of variance for the TSS response of the variables taken as a response for industry A, with only the Total Soluble Solids (° Brix) response variable proving statistically significant over storage time.

Table 4.4.1 Estimation of regression coefficients

	SQ	GL	MQ	F-test
Regression	2,090313	3	0,69677	44,636
Waste	0,0468	3	0,01561	
Lack of adjustment	0,00016	1		
Pure error	0,05	2		
Total	2,14	6		
R^2	97,809	-		
Ftabeled 0.95, 3, 3				9,18

The data was analyzed considering pure error and a 95% confidence limit. It can be seen that the coefficient of determination, R^2 = 97.80 %, with the equation for the model, TSS (°Brix) 5.20 + **0.058 Xi** + 0.044 X2 - 0.**008** x_1x_2. Response = (B_0 + B1X1 +B2X2 + $B_4X_1X_2$), where x_1 is storage temperature and X2 is time.

The model proved to be statistically significant for the TSS response (° Brix), because according to Rodrigues and Iemma, (2005) the calculated F has to be higher than the tabulated F and the ratio of the calculated F to the tabulated F greater than 1. In this case the ratio was 4.86, showing that the model is not only significant but also considered predictive with a ratio greater than 1.

The importance of the main effects together with the interaction effects can be better visualized by the Pareto chart. The magnitude of the static effects is represented by the columns while the line perpendicular to the columns represents the magnitude of the effects with statistical significance for p = 0.05.

Figure 4.1 shows the Pareto diagram representing the independent variables, time and temperature. It can be seen that the interaction of time and temperature negatively influenced the total soluble solids values, as did the isolated effect of temperature, while time alone had no influence on total soluble solids.

Figure 4.1- Pareto diagram - Effect of TSS, temperature and refrigerated storage time.

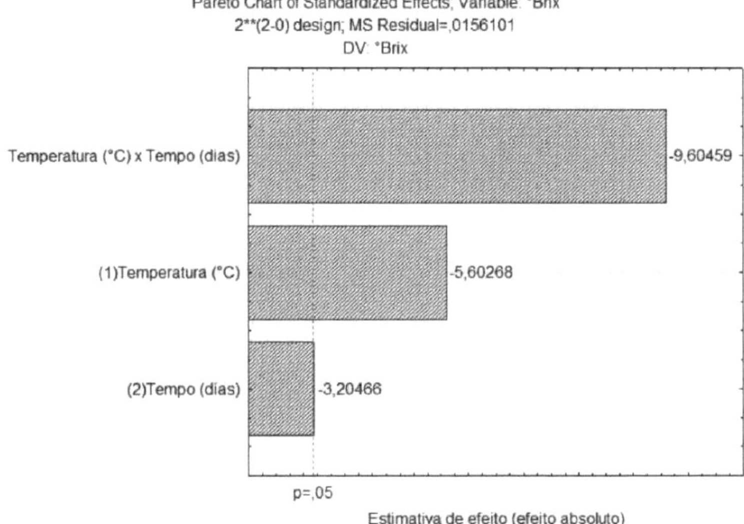

The effect of temperature and time on the TSS content of coconut water stored for 30 days is shown in Figure 4.2 as a response surface.

It can be seen that higher temperatures led to an increase in these solids until approximately the fifth day of storage, while lower temperatures led to an increase as storage time progressed.

In the binomial of time and temperature, it can be seen that by increasing the temperature and time, there was a reduction in the total soluble solids content.

Average values were obtained (5.5 ° Brix), range 4 to 6 °C and 0 to 10 days. This is due to the metabolic processes and microbiological damage that accelerate, thus causing the Brix to drop as the temperature rises.

Figure -4.2 Response surface contour plot for the variable TSS (°Brix)

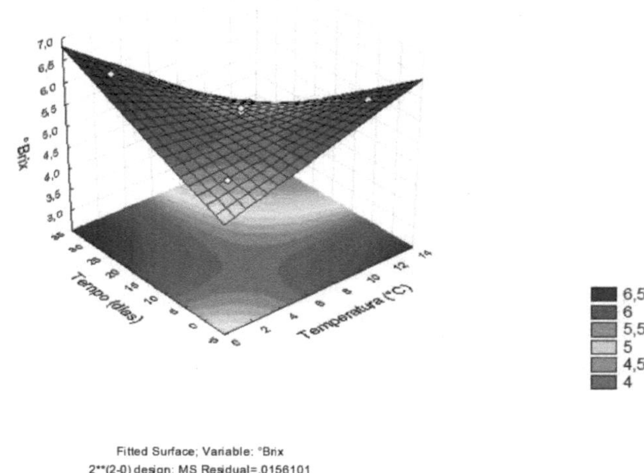

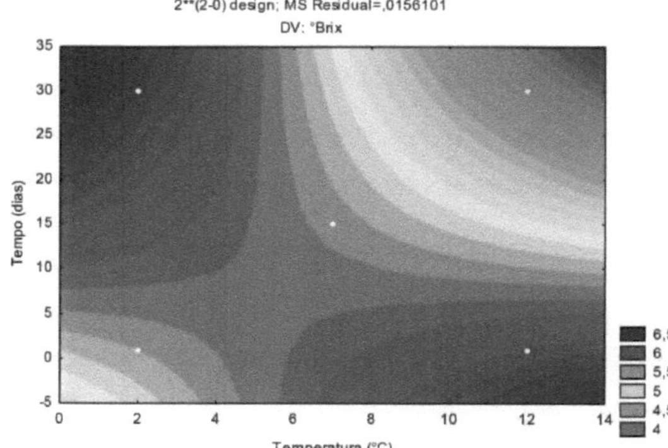

For industry B, the response variables conductivity and viscosity were the ones with statistically significant models.

Table 4.4.2 shows the results of the analysis of variance (ANOVA) for the response variable Conductivity for the coconut water analyzed from industry B.

Table -4.4.2 Estimation of regression coefficients

	SQ	GL	MQ	F-test
Regression	0,170313	3	0,05677	24,9352
Waste	0,0068	3	0,00228	

Lack of adjustment	0,00016	1
Pure error	0,01	2
Total	0,18	6
R^2	96,144	-
Ftabeled 0.95, 3, 3		9,18

The data was analyzed considering the pure error and a confidence limit of 95%. It can be seen that the coefficient of determination, R^2 = 96.14 %, with the following equation for Conductivity 4.342 + **0.010 Xi** - 0.044 X2 - 0.018 X2 + **O,0006x1x2**. Response = (Bo + B1X1 +B2X2 + B4X1X2), where X1 is storage temperature and X2 is time.

The model was statistically significant for the response Conductivity of coconut water for industry B, as the calculated F was greater than the tabulated F, with a value of 24.93 and the ratio between the calculated F and the tabulated F was 2.71, showing that, as well as being significant, the model was predictive with this ratio being greater than 1.

Figure 4.4 shows the Pareto diagram representing the independent variables time and temperature. It can be seen that time alone had a negative influence on the electrical conductivity of coconut water, i.e. as the storage time progressed, the electrical conductivity decreased. The interaction between time and temperature and temperature alone did not.

Figure 4.3- Pareto diagram - Effect of conductivity, temperature and refrigerated storage time.

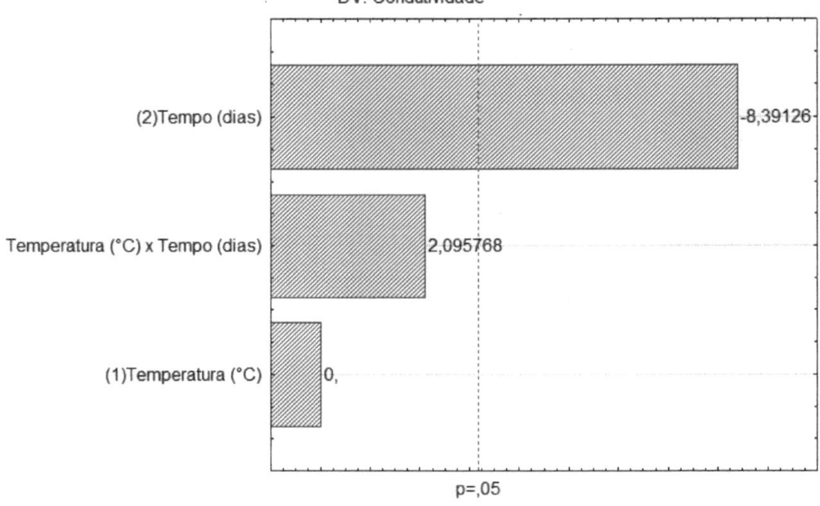

Conductivity represents the levels of ions present in coconut water, such as sodium, potassium, calcium and magnesium ions (PINHEIRO, 2009).

Figure 4.4 shows the response surface diagram and the level curve corresponding to the fitted model, which establishes the variation in conductivity as a function of storage time and temperature for processed coconut water.

Figure 4.4 Response surface contour plot for the response variable Conductivity

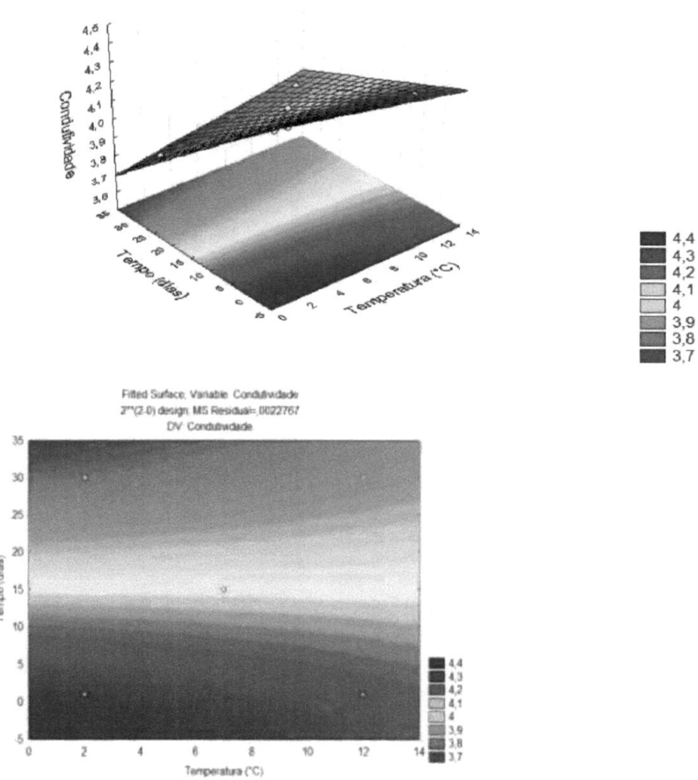

Higher electrical conductivity values (4.2 to 4.4) were obtained up to the fifth day of storage, at all the temperatures studied, while the lowest values for this variable were observed from the fifteenth day of storage (4-3.7). In general, there was a decline in electrical conductivity as storage time progressed at all the temperatures studied.

Table 4.4.3 shows the analysis of variance for the response variable viscosity of coconut water from industry B.. The data was analyzed considering the pure error and a 95% confidence limit.

Table 4.4.3 Estimation of regression coefficients

	SQ	GL	MQ	F-test

Regression	0,136657	3	0,04555	125,87
Waste	0,0011	3	0,00036	
Lack of adjustment	0,00109	1		
Pure error	0,00	2		
Total	0,14	6		
R^2	99,212	-		
Ftabeled 0.95, 3, 3				9,18

It can be seen that the coefficient of determination was R^2 = 99.21 %, indicating that the model is statistically significant with the equation 0.327 - 0.**001 Xi** + 0.0014 X2 + 0.**0011 x1, x2,**viscosity. Response = (B_0 + B_1X_1 + B2X2 + $B_4X_1X_2$)X1 is storage temperature and X2 time.

It can be seen that the interaction of temperature and time influenced the viscosity of the stored coconut waters, with the advance of storage time and higher temperatures there was an increase in viscosity. Time and temperature alone also influenced the increase in this variable.

Figura 4.5 Pareto diagram - Effect of viscosity, temperature and refrigerated storage time.

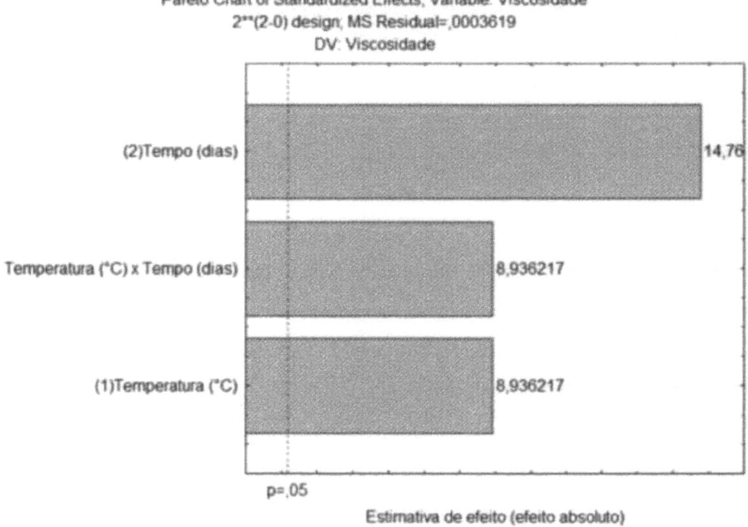

Figure 4.6 shows the response surface and the level curve corresponding to the adjusted model that establishes the variation in viscosity as a function of time and storage temperature for the coconut water processed in industry B.

Figura 4.6 - Contour plot of the response surface for the viscosity variable

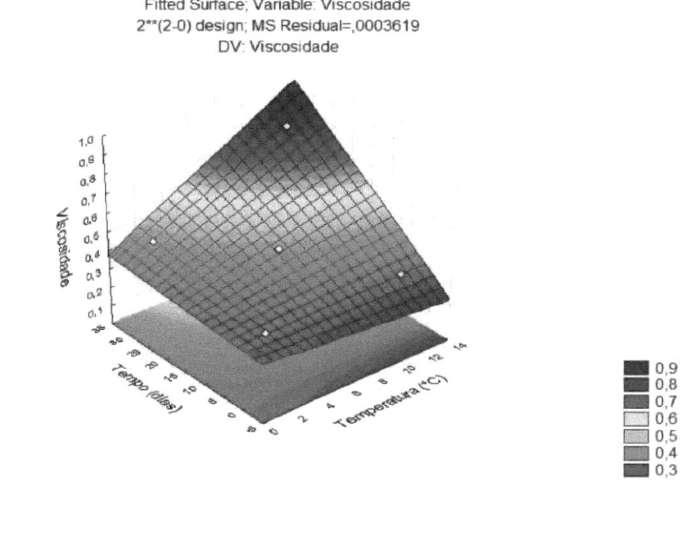

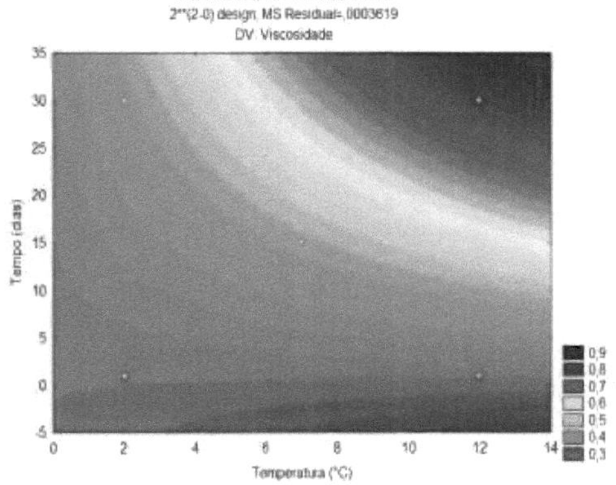

The results obtained in the study show that the viscosity of coconut water varied little up to the 10th day of storage at all the temperatures studied, but from this period onwards there was an increase in this variable at temperatures above 4 °C. In general, higher viscosity values were observed at temperatures above 8 °C and from 20 days onwards.

SHAMES (1999) comments that viscosity is directly proportional to the force of attraction between molecules. As the temperature rises, this force of attraction decreases, which also reduces viscosity. Thus, in liquids, viscosity decreases with increasing temperature.

By analyzing the response surface graphs for the variables analyzed, we can conclude that refrigerated coconut water does not have a shelf life of thirty days without sensory losses. Refrigerated coconut water should be stored at low temperatures for a maximum of ten days, which is suitable for human consumption.

4.5 Microbiological results

Tables 4.5.1 and 4.5.2 show the microbiological analyses of the dwarf green coconut water sold by the industries after 30 days' storage under refrigeration.

Table 4.5.1 Results of the microbiological analysis of green dwarf coconut water marketed by industry A.

| Experiments | Variables | | Answers | | | | |
| | Temperature (°C) | Time (day) | | | | | |
	X1	X2	Coliforms 35°	Coliforms 45°	Salmonella	Molds and yeasts	Psychrotrophic
1	2	1	$1,5X10^2$	7	Absent	5,05x102	6,7x102
2	12	1	$1,5x10^2$	7	Absent	7,2X102	$1,5x10^3$
3	2	30	7	< 3	Absent	$4,0X10^5$	< 7
4	12	30	7	< 3	Absent	$7,6x10^5$	$> 10^3$
5	7	15	7,5x10	< 3	Absent	$5,6x10^4$	$4,2x10^4$
6	7	15	7,5x10	< 3	Absent	$2,2X10^4$	$1,9x10^3$
7	7	15	1,5x102	< 3	Absent	$2,2X10^4$	$2,5x10^5$

In Table 4.5.1, the samples of anao verde coconut water sold by industry A showed total coliform results ranging from 7 NMP/mL to 1.5 x 10^2 NMP/mL and < 3 NMP/mL to 7 NMP/mL for thermotolerant coliforms, the values varying over the thirty days of storage. Almadas et al. (2009), analyzing bottled coconut water sold in Currais Novos, RN, found similar data regarding thermotolerant coliform contamination. Penha et al. (2005), when studying the physical, chemical, microbiological and sensory characterization of coconut water bottled using the aseptic process, found that there were no thermotolerant coliforms. This fact can be explained by the process that coconut water undergoes, with the use of heat treatment and the effective elimination of pathogenic microbiota. Coliforms are a group of enterobacteria present in feces and in the environment, such as soil and the surfaces of plants, animals and utensils. Their detection in food is used as a reliable indicator of the hygienic conditions of the product and the presence of enteropathogens. According to Brasil (2009), which establishes the microbiological standards of coconut water for consumption, the samples analyzed are in disagreement with the microbiological standards of Normative Instruction No. 27, of July 22, 2009, which establishes results for coliforms at 45 °C (thermotolerant coliforms) of less than 1NMP/mL.

Oliveira (1992) points out that many Brazilian fruit and vegetables are not only irrigated with water contaminated by pesticides and fecal material, but also with human waste. This is why the consumption of raw fruit and vegetables is a strong means of transmitting infectious and parasitic diseases to the population.

Salmonella sp was not detected in any of the coconut water samples analyzed, in accordance with current legislation (BRASIL, 2009), which establishes its absence in 25 mL(g) of the sample. Similar results were found by Carvalho et al (2012) and Silva et al (2009), who assessed the physical, chemical and microbiological quality of coconut water bottled in Teresina, Piaui, and concluded that 100% of the samples analyzed did not contain *Salmonella* sp. According to Alves Filho (2003), the lack of hygienic care during the production and/or handling of fruit increases the risk of contamination by *Salmonella* sp, a microorganism that causes diseases that can lead to death.

In the present study, high bacterial counts were found in the samples of coconut water from industry A, under refrigeration, thus indicating the use of contaminated raw materials, inadequate processing, or even unsatisfactory storage, related to the time and temperature that the samples were kept.

Molds and yeasts are responsible for the deterioration of fruit juices, frozen, dehydrated and preserved foods when stored in unsuitable conditions. Their growth is favored by low acidity and high water activity. As in the case of coconut water, which has a very slight acidity of around 0.22% in normal solution, a pH very close to 5.0 and a high humidity rate of over 93% (FRANCO ; LANDGRAF, 2003).

The values found for moulds and yeasts ranged from 5.05×10^2 UFC/mL to 7.6x105 UFC/mL, which are considered high, characterizing coconut water as unsuitable for consumption when compared to current legislation (BRASIL 2009), which sets a maximum limit of 20 UFC/mL. Spoilage is only noticeable when mold growth is visible or the food has a high number of yeasts, and spoilage by yeasts is usually not harmful to health (SANTOS and RIBEIRO, 2006).

According to Almadas, et al. (2009), when analyzing bottled coconut water sold in Currais Novos, RN, the presence of moulds and yeasts was found to be lower than in this study, ranging from 6.0 x 10 CFU/Ml to 1.7×10^5 , suggesting the need for improvements during the bottling process of chilled green coconut water.

As for psychrotrophic microorganisms, these are microorganisms that can grow at a temperature between 0 °C and 7 °C, they have the ability to form visible colonies (or turbidity) within 7 to 10 days in this temperature range and their optimum growth temperature is between 20 °C and 30 °C and the maximum is 42 °C, however certain molds and psychotrophic bacteria can still grow at temperatures as high as 58 °C (JAY et al., 2005).

Industry A had a count of psychrotrophic microorganisms ranging from < 7 CFU/mL to 2.5×10^5 CFU/mL at all the times of analysis studied. Although current legislation does not establish standards for psychrotrophic bacteria, it is known that psychotrophic microorganisms are capable of multiplying at low temperatures, although their multiplication occurs more slowly under refrigeration. According to Sivasankar (2004), the deterioration that occurs in food subjected to temperatures above 10 °C is about twice as fast as when subjected to temperatures between 0 °C and 5 °C.

The effectiveness of refrigeration is based on reducing the activity of the microorganisms present in food, which slows down the degradation of its components and consequently increases the shelf life of the products. This time depends on the nature of the food itself, but also on its initial contamination, which the lower it is, the longer its shelf life will be under identical conditions of preservation (BATISTA; ANTUNES 2005).

According to the Brazilian Association of Food Quality Professionals, all people who have contact with any stage of food processing must be trained and made aware to practice hygiene and safety measures in order to protect food from chemical, physical and microbiological contamination (BADARÓ, 2007).

Table 4.5.2 Results of the microbiological analysis of green dwarf coconut water marketed by industry B.

| Experiments | Variables | | Answers | | | | |
| | Temperature (°C) | Time (day) | Coliforms 35° | Coliforms 45° | Salmonella | Molds and yeasts | Psychrotrophic |
	X1	X2					
1	2	1	2,1X10	7	Absent	$2,2 \times 10^3$	3,2x103
2	12	1	2,8X10	7	Absent	$3,7 \times 10^3$	1,3x10
3	2	30	<3	<3	Absent	<7	<7
4	12	30	<3	<3	Absent	$5,6 \times 10^5$	>103
5	7	15	<3	<3	Absent	$6,1X10^3$	3,7x10

| 6 | 7 | 15 | <3 | <3 | Absent | $3,4 \times 10^5$ | 9,6x103 |
| 7 | 7 | 15 | <3 | <3 | Absent | $4,5 \times 10^4$ | 2,1x10 |

Table 4.5.2 shows the results of the microbiological evaluations carried out by the research with industry B during the thirty days of storage under refrigeration, which show total coliform contamination ranging from < 3 NMP/mL to 2.8x10 NMP/mL and thermotolerant coliforms ranging from < 3 NMP/mL to 7 x10 NMP/mL, which is outside the standards established by current legislation (BRASIL 2009), which establishes microbiological standards for coconut water for consumption with results of thermotolerant coliforms (fecal coliforms) of less than 1 NMP/mL. Santos et al.(2013), when studying the microbiological evaluation of coconut water sold by street vendors in Juazeiro do Norte CE, found results ranging from < 3 NMP/mL to 1.1×10^3 NMP/mL. The presence of faecal coliforms indicates that there has been direct or indirect contact with human or warm-blooded animal faeces in the product or during the process, characterizing a deficiency in the handling or poor hygiene of the raw material.

However, Silva Júnior (2005) reports that these microorganisms are present in the environment. Therefore, the presence of total bacteria is an indicator of the hygienic and sanitary conditions of the preparation area. The evaluation of this count is commonly used as an indication of quality in food production.

With regard to *Salmonella* sp in the samples analyzed, there was no presence of *Salmonella* sp, in accordance with current legislation (BRASIL 2009), which establishes its absence in 25ml (g). Santos et al. (2013), when studying the microbiological evaluation of coconut water sold by street vendors in Juazeiro do Norte CE, found no *Salmonella* sp in their studies and Walter et al. (2009) when studying the modeling of bacterial growth in fresh green coconut, concluded that the drink is conducive to the growth and survival of these bacteria and that refrigeration at temperatures between 4 and 10 °C delays but does not inhibit growth. In 1999, the US Centers for *Disease Control* (CDC, 2006) reported an epidemic of 207 confirmed cases of diarrhea caused by drinking unpasteurized orange juice. In just one month, hundreds of inhabitants of 15 American states and two Canadian provinces had consumed a drink contaminated with *Salmonella* sp. This forced government health authorities to carry out a wide-ranging emergency action to report the cases and recall the product available in supermarkets and restaurants in both countries. The *Salmonella* sp bacterium alone has more than 2,400 serotypes that are pathogenic to humans. The action of these pathogens depends on the poor hygiene conditions of the environment and the susceptibility of the human host. This has had serious implications for human health.

As for molds and yeasts, which are not pathogenic microorganisms, but deteriorating ones, the results showed a range from < 7 CFU/mL to 5.6×10^5 CFU/mL. This is not in accordance with current legislation (BRASIL 2009), which establishes a maximum limit of 20 CFU/ML, so it can be concluded that the coconut waters analyzed have a high potential for contamination by spoilage microorganisms.

It is known that green coconut water still inside the fruit is sterile, but deficiencies in the hygiene and sanitation aspects arising from the stages of washing and handling the fruit, as well as cleaning and sanitizing the storage and preservation equipment, can spread contamination and the proliferation of microorganisms, making the product unfit for consumption.

Santos et al. (2001) found results indicating contamination in refrigerated coconut water sold on the streets and in commercial outlets in the city of Aracaju. Hoffman et al. (2002) found that 25% of the different samples of coconut water sold in the municipality of Sao José do Rio Preto - SP did not comply with current Brazilian legislation.

The coconut water packaging process in factories A and B is done by hand and is therefore more susceptible to contamination. This does not prevent a beverage such as coconut water from being packaged in flexible containers, as long as they are made from laminated films composed of several different materials: polyester on the outside, aluminum (oxygen barrier), aluminum cans, polypropylene cups, laminated cartons and glass and PET bottles as long as they are sanitized and filled aseptically (PETRUS, 2005).

Looking at table 4.5.2 for industry B, it can be seen that *the* presence of psychrotrophic bacteria ranges from < 7 CFU/ mL to 9.6×10^3 CFU/ Ml, which is close to industry A. According to a study by Garbutt (1997), refrigeration chambers oscillate between a temperature of 4 °C and 7 °C, and at these temperatures psychrotrophic microorganisms can proliferate, albeit at a slow rate, making the product potentially dangerous for consumer health.

When analyzing the factories, we can conclude that the results obtained for the microbiological analyses showed unsatisfactory microbiological quality, as determined by the resolution of the Ministry of Agriculture and Livestock and Supply (BRASIL, 2009), except for *salmonella sp,* which was absent in all the analyses, in accordance with Brazilian legislation. We can see flaws in the bottling and/or preservation process of coconut water bottled and refrigerated by the industries.

According to Batista; Antunes (2005), the main characteristic of perishable foods is the fact that they deteriorate easily, and the deterioration process can begin at the time of purchase, or even before, although these products must be stored at low temperatures.

Refrigeration is therefore an important barrier for controlling microbial growth, but failures in the cold chain are common. It is necessary for the method used to preserve the product to be based on more than one control factor Lika; Jevsnik (2006) established that refrigerated storage for perishable foods should be set at temperatures $\leq 5\ °C$.

4.6 Sensory analysis

In the process of developing and improving products, determining acceptance is extremely important. Acceptance tests require a large number of participants who represent the population of current or potential consumers of the products Sheid (2001). In this case, the sensory analysis of the present study was conducted with the aim of verifying consumer acceptance of the general characteristics of green dwarf coconut water produced commercially by industries in the sertao of Paraiba and Cearà.

The average acceptance of the attributes of appearance, aroma, flavor, purchasing attitude and product preference of each of the coconut water samples studied by the Tukey test corresponded to experiments 1 and 2 on the day they were manufactured.

The sensory study could only be carried out in the experiments mentioned above, which could not be followed in experiments 3, 4, 5, 6, 7, corresponding to temperatures of $7\ °C$, $2\ °C$ and $12\ °C$, at 15 and 30 days of storage, as they were unfit for consumption, cloudy, with perceptible changes in the organoleptic level. At 15 days of storage at the temperatures mentioned above, 70% of the coconut water cups stored were cloudy and produced gas. After 30 days of storage, all the glasses were unsatisfactory for consumption, as can be seen in figures 4.7 and 4.8.

Figura 4.7 - Storage of coconut water processed by the industries with 15 days of refrigerated storage at $7\ °C$. Source: author

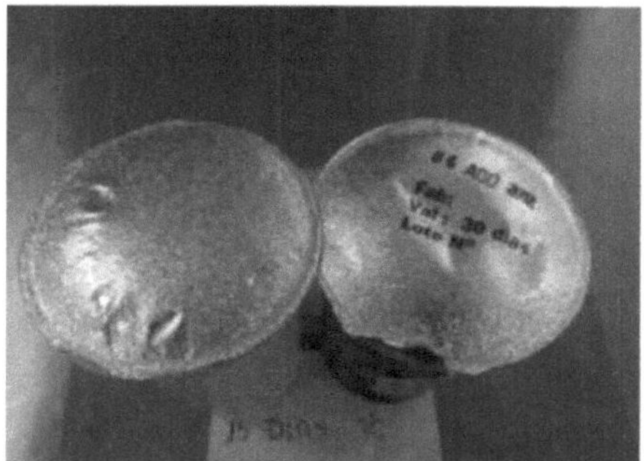

Figura 4.8 - Storage of coconut water processed by the industries with 30 days of refrigerated storage at 12°C. Source: author

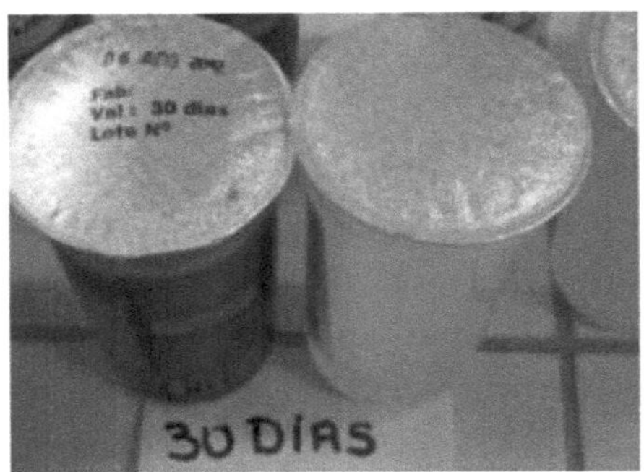

Tables 4.6.1 and 4.6.2 show the approximate average scores given by the 40 tasters for the attributes of appearance, taste and aroma, flavor, purchasing attitude and product preference of the coconut water filled and sold by the industries on the day it was manufactured.

Table 4.6.1 shows the approximate average scores given by the 40 tasters for the appearance, taste and aroma attributes of the coconut water packaged and marketed by the industries.

Table 4.6.1 - Average scores given by the tasters for the sensory attributes of appearance, flavor and aroma of the green dwarf coconut water produced commercially by the industries in experiments 1 and 2.

Experiments	Attributes					
	Appearance (ind. A)	Appearance (ind. B)	Flavor (ind. A)	Flavor (ind. B)	Aroma (ind. A)	Aroma (ind. B)
Exp.1	7.87a	7.27a	8.07a	5.67a	7.97a	6.90a
Exp.2	7.77a	7.07a	7.72a	5.77a	7.35b	6.10b
DMS	0.29976	0.62718	0.39296	0.63323	0.47316	0.70440
MG	7.82	7.17	7.90	5.72	7.66	6.50
CV(%)	8.60	19.63	11.17	24.84	13.87	24.34

MSD - Minimum significant deviation; MG - General average; CV - Coefficient of variation.

Means followed by the same small letter in the columns do not differ statistically by Tukey's test at 5% probability. Scores: 1= disliked very much; 2= disliked very much; 3= disliked moderately; 4= disliked slightly; 5= neither liked nor disliked; 6= liked slightly; 7= liked moderately; 8= liked very much; 9= liked extremely much.

The results of the sensory evaluation, shown in Table 4.6.1, for the refrigerated coconut water product for industry A, in terms of appearance and taste, were approximately 8 "I liked it very much", classified by the nine-point structured hedonic scale, in both experiments. For industry B, they were approximately 7 "I liked it moderately" for the appearance attribute and 6 "I liked it slightly" for the flavor attribute, in both experiments, which showed no significant differences by Tukey's test at a 5% probability level between the experiments evaluated, demonstrating the non-interference of storage temperatures for green dwarf coconut water on the day of its manufacture.

Looking at table 4.6.1, which refers to the aroma attribute, industry A gave a score of approximately 8 "I liked it very much" in experiment 1 and 7 "I liked it moderately" in experiment 2. For industry B, the scores given were approximately 7 "I liked it moderately" in experiment 1, and 6 "I liked it slightly" in experiment 2. It is possible to see that they differed statistically, showing significant differences by Tukey's test at a 5% probability level between the experiments evaluated.

According to Nogueira et al. (2004), who studied the sensory evaluation of fresh and processed coconut water, they found that fresh coconut water was more acceptable to tasters (33%) than processed coconut water (22%), although it is more convenient to buy and store.

Barros et al. (1998), who studied the preservation of coconut water by refrigeration, found that á coconut water was suitable for consumption after being bottled and stored at a temperature of 10 °C for 48 hours, even though there was a change in the color of the product after 48 hours of storage. However, this change did not affect the product's acceptability.

Comparing the contrasts between the samples, it can be seen that industry A obtained better results in the appearance, flavor and aroma attributes for experiment 1 when compared to the same experiment by industry B, which had lower scores.

Table 4.6.2 - Average scores given by the tasters for the sensory attributes of purchasing attitude, product preference and green dwarf coconut water produced commercially by the industry in experiments 1 and 2.

Experiments	Attributes			
	Attitude of purchase (A)	Buying attitude(B)	Product preference (A)	Product preference (B)
Exp.1	4.62a	3.32a	1.47a	3.12b
Exp.2	4.65a	3.00a	1.65a	3.77a
DMS	0.26904	0.54014	0.26234	0.22178
MG	4.63	3.16	1.56	3.45
CV(%)	13.03	38,36	37,71	14,44

MSD - Minimum significant deviation; MG - General average; CV - Coefficient of variation.

Means followed by the same small letter in the columns do not differ statistically by the Tukey test at 5% probability.

Table 4.6.2 shows that for all the attributes evaluated, the coconut water samples from industry A obtained the best results. For the attitude to purchase attribute, in experiment 1 and experiment 2, the values obtained were close to 5, which on the five-point structured hedonic scale would mean I would certainly buy. There were no significant differences between the storage temperatures under study. With lower scores for industry B, the scores given for purchasing attitude were close to 3, which on the scale would be maybe I would buy/ maybe I wouldn't buy, with no significant differences by Tukey's test at a 5% probability level between the experiments evaluated.

For the results shown in table 4.6.2, the approximate averages of the product preference attributes, industry A obtained values close to 1, which on the scale would be first for experiment 1, and 2, which on the scale would be second for experiment 2.

Industry B obtained a lower value for the product preference attribute, with an average of approximately 3, which on the scale would be third place for experiment 1, and 4, fourth place for experiment 2. One of the major concerns of industries when developing a new product is to verify the consumer's purchasing intention (SANTANA et al., 2006).

The results of the purchase intention test showed that the term "would certainly buy" on the scale obtained a value of approximately 5. There were no significant differences between the storage temperatures under study. In terms of product preference, there were no differences, with the best average given by the tasters to experiment 1, which on the scale under study would be first, and experiment 2 second.

5. CONCLUSIONS

• With storage time, there was a decrease in the pH, vitamin C, conductivity and total soluble solids values for both factories, with the total soluble solids values being in line with what is allowed by BRASIL (2009) legislation, except for experiment 4. The pH of all the experiments does not comply with legislation. There were simultaneous increases for the industries in the analyzed parameters of turbidity, viscosity and total titratable acidity.

• The samples analyzed microbiologically for both industries do not comply with the standard established by current legislation for thermotolerant coliforms, molds and yeasts. No *Salmonella sp.* values were found in any of the samples analyzed.

• Sensorially, the coconut water was already unsuitable for human consumption after 15 days of storage, showing altered organoleptic characteristics; Industry A showed better values for all the sensory attributes evaluated;

• The time x temperature binomial considerably influenced the parameters analyzed during storage;

• It is essential to implement and monitor Good Manufacturing Practices in places where this type of food is produced or handled, since coconut water is a major vehicle for microbial contamination as it is a perishable product.

6. Suggestion

It would be technologically important for the method of preservation by refrigeration if Brazil (2009), in its ordinance governing the standards and quality of bottled and refrigerated coconut water, evaluated another microbiological parameter for natural coconut water, as it is a perishable product and susceptible to the proliferation of psychrotrophic bacteria due to its storage at low temperatures. This provides another parameter for assessing the quality of refrigerated coconut water for packaging and marketing.

7. BIBLIOGRAPHICAL REFERENCES

ABREU, F A. P., Processing of coconut water preservation by combined methods. DEINPI/CE Patent N° 000129, September, 1999.

ABREU, L. F. 2005. **Evaluation and adaptation of an aseptic system for obtaining coconut water (*Cocus nucifera L.*)** in plastic packaging. Thesis (Doctorate in Food Technology) - State University of Campinas, Campinas.

ABREU, Fernando. TRADE, EXTERIOR... Available at http://www.sfiec.org.br/noticias/export-aguacoco. Accessed on January 3, 2008.

NATIONAL HEALTH SURVEILLANCE AGENCY (ANVISA). Food. Good Practices. Available at: <http://www.anvisa.gov.br/alimentos/bpf.htm>. Accessed on: 17/05/2013.

ALVES FILHO, M. Research investigates risks and benefits of foods and nutrients. **Jornal da Unicamp**. Ed.211-5 to May 11, 2003. 6-7p.

ALMADAS, DANTAS; SILVA. **Holos,** Year 25, Vol. 3. Page 34. Microbiological quality of coconut water sold in the municipality of Currais Novos, RN (2009).

ANDRADE, M. V. V. et al. Microbiological evaluation of coconut water. In: SIMPÒSIO MINEIRO DE MICROBIOLOGIA DOS ALIMENTOS, 3., 2008, Viçosa, MG. **Proceedings...** Viçosa, MG: UFV, 2008.

AGRICULTURE 21. Enfoques: Nueva bebida para el porte: agua de coco. **FAO magazine.** Available at:<www.fao.org/ag/esp/revista/. Accessed on: September 22, 2003.

ARAGÂO, W. M.; ISBERNER, I. V.; CRUZ, E. M. O. **Agua-de-coco.** Aracaju: Embrapa CPATC/ Tabuleiros Costeiros, 2001 (Documents Series 24).

ARAGÂO, W.M; CRUZ, E. M.; HELVÉCIO, J. S. Morphological characterization of fruit and coconut water chemistry in dwarf coconut cultivars. **Agrotópica**, v. 13, n. 2, p. 49-58, 2001.

ARAGÂO, W.M. A importância do coqueiro-anao verde. Articles EMBRAPA **Coletânea rumos & debates**, 20/06/2000.

ARAGAO, W. M.; RESENDE, J. M.; CRUZ, E. M. de O. ; REIS, C. dos S.; JUNIOR, O. J.S.; ALENCAR, J. A. de.; MAREIRA, W. A.; PAULA, F. R. de.; FILHO, J. M. P. L. Coconut fruits for natural consumption. In: ARAGAO, W. M. et al. (Ed) **Coconut post-harvest.** Brasilia: Embrapa Informaçoes Tecnològica; Aracaju: Embrapa Tabuleiros Costeiros, 2002. Chap. 3, p. 19-25. (Frutas do Brasil, 29).

ARAGÂO, W.M. (Ed.) et al. Coconut: Post-harvest. **Série frutas do Brasil**, 29. Brasilia: Embrapa

Informaçao Tecnològica, 2002. 76p.

ARAGÂO, F.B.; LOIOLA, C.M.; CAMBUI, E. V. F.; ARAGAO, W. M. **Coconut water production and green coconut cultivars.** Aracaju: EMBRAPA, 2005.p.1-2. (Technical Communication, 42)

ARAGÂO, W. M. et al. Coconut **fruit for natural consumption.** In: ARAGÂO, W. M. Coconut post-harvest. Brasilia: Embrapa informaçao tecnològica, 2002. 76 p.

AROUCHA, E. M. M. **Evaluation of the main physical and chemical characteristics of the liquid and solid endosperm of the Ana (Cocos nucifera L.) Green and Red coconut cultivars at different stages of ripeness.** 2000. 86f. (Master's Degree in Plant Production). Universidade Estadual do Norte Fluminense, Rio de Janeiro.

AROUCHA, E. M. M., SOUZA, C. L. M.; AROUCHA, M. C. M.; VIANNI, R. Physical and chemical characteristics of Anao verde and Anao vermelho coconut water at different stages of ripeness. **Caatinga**, Mossorò, v. 18, n. 2, p. 82-87, 2005.

ARRIOLA, M. C. de; CALZADA, J. F. de; MENCHU, J. F.; ROLZ, C.; GARCIA, R.; CABRERA, S. De. **Papaya.** In: Tropical and subtropical fruits. Westport: AVI, p. 316-340, 1980

ASSIS, J.S.; RESENDE, J.M.; SILVA, F.O.; SANTOS, C. R., NUNES, F. Técnicas para colheita e pós colheita do coco-verde. Petrolina: Embrapa-CPATSA, 2000.3p. Embrapa-CPATSA (technical communiqué, 153/2).

ASSIS, O. B. G. de. Edible **protective coatings on fruit**: an emerging technology.Disponivel:<http://www.todafruta.com.br/portal/icNoticia7Aberta.asp?id Noticia= 14349>. Accessed on: Jan. 24, 2012.

AZOUBEL, P.M.; MURR, F. EX. Optimization of osmotic dehydration of cashew apple (Anacardiumoccidentale L.) in sugar solutions. **Food Science and Technology International,** London, v.9, n.6, p.427-433, 2003.

BADARÓ, A.C.L. **Boas praticas para serviços de Alimentaçâo: um estudo em restaurantes comerciais do município de Ipatinga, Minas Gerais. 2007.** 172p. Dissertation (Master's Degree in Nutritional Sciences) - Department of Nutrition and Health, Federal University of Viçosa. 2007.

BARROS NETO, B., SCARMINIO, I. S., BRUNS, R. E. **Planejamento e Otimizaçâo de Experimentos.** 2 ed., Campinas, SP, Editora da Unicamp, 1995, 299 p.

BARROS, P. R. et al. Preservation of coconut water by refrigeration. **B. CEPPA**, Curitiba, v.16, n. 1, p. 1-12, jan./ jun.1988.

BATISTA, P. ANTUNES, C. (2005). **Hygiene and Food Safety in Catering.** Vol.

II (1^a ed.). Forvisao - Consultadoria em Formaçao Integrada. Guimaraes, Portugal.

BHATTACHARYYA, A.; BHATTACHARYYA, N. Coconut in nutrition .Indian **Journal of Nutrition and Dietetics**, v.39, n.3, p. 132-142, 2002.

BRAZIL. Ministry of Agriculture, Livestock and Supply. **Normative Instruction No. 39, of July 22, 2009.** Approves the Technical Regulation for establishing the identity and quality of coconut water.

BRAZIL. Ministry of Agriculture, Livestock and Supply. Normative Instruction No. 62, of August 26, 2003. Official Analytical Methods for Microbiological Analyses for the Control of Products of Animal Origin and Water. **Official Gazette of the Federative Republic of Brazil,** Brasilia, 18/09/2003, Section 1, page 14.

BRAZIL. Ministry of Health. National Health Surveillance Agency. **Chemical and physico-chemical methods for food analysis.** Brasilia: Ministry of Health, 2008. 1017 p.

BRAZIL. Ministry of National Integration. Secretariat for Water Infrastructure. Department of Special Projects. Coco-Green. **Frutiséries**, Brasilia, n. 3, 4p, 2000.

BRITO M.A.V.P. Qualidade do leite a partir de detalhes. BaldeBranco, October, p. 66-74, 2001.

BROUCHT, W.J.; TAN, G.Storage condition and ripening of custard apple (Annonasquamosa L.).**ScientiaHorticulturae,** Amsterdam, v. 10, n. 1 p. 73-82, 1979

BENASSI, A. C. **Biometric, chemical and sensory characterization of Ana Verde coconut fruits.** Thesis (Doctorate in Agronomy). Paulista State University "Julio de Mesquita Filho". Faculty of Agricultural and Veterinary Sciences, Jaboticabal. 2006. 98p.

CABRAL, L. M. C.; Lourdes, M. C. C.; Edmar, M. P.; Virginia, M. M. Agua de coco verde refrigerada / Brasilia, DF: Embrapa Informaçao Tecnològica, **Agroindùstria familiar.** 2005.p 34.

CAETANO, V. C; SALTIVI, D. A., PASTERNAK, J. Outbreak of salmonellosis by enteric salmonella in health professionals, caused by food consumed at a New Year's party held inside the intensive care unit. **Einstein.** 2004; 2(1): 33-5p.

CAMPOS, C. F.; SOUZA, P. E. A. COELHO, V.; GLORIA, M. B. A. Chemical composition, enzyme activity and effect of enzyme inactivation on flavour quality of green coconut water. Journal of Food Processing and Preservation, 20, 487-500.1996.

CARNEIRO, WMA. Public policy and income in family farming: the influence of the Cariri Cearense Agribusiness Development Pole. In: SOBER ANNUAL CONGRESS, 45. **Full article.** Londrina: SOBER (CD-ROM). 2007.

CARVALHO, J. M. **Drinks based on coconut water and cashew juice: processing and stability.** 2005. 107f. Dissertation (Master's Degree in Food Technology) - Federal University of Ceara, Fortaleza, 2005.

CARVALHO, J. M.; MAIA, G. A.; BRITO E. S.; CRISÓSTOMO, L.A.; RODRIGUES, S. Mineral composition of a mixed drink based on coconut water and clarified cashew juice. B. CEPPA, Curitiba, v.24, n.1, p.1-12, 2006.

CARVALHO, L. R.; PINHEIRO, B. E. C.; PEREIRA, S. R.; BORGES, M. S.; MAGALHAES, J. T. Antimicrobial-resistant bacteria in samples of coconut water sold in Itabuna, Bahia. Revista Baiana de Saùde Pùblica, v.36, p. 751- 763, 2012.

COSTA, L.M.C.; MAIA, G.A.; COSTA, J. C, et al .Evaluation of coconut water obtained by different preservation methods.**CiênciaAgrotécnicas**. Lavras, v.29, n, 6, p. 1239-1247, nov./ dez.,2006.

CUENCA, M. A. G.; RESENDE, J. M.; SAGGIN JÛNIOR, O. J. et al. The Brazilian coconut market: current situation and prospects. In: ARAGÂO, W. M. Coconut: post-harvest. Brasilia, DF: Embrapa. **Informaçâo Tecnològica**, 2002. p. 11-18.

CUENCA, M. A. G. Coconut cultivation. Embrapa Tabuleiros Costeiros, Sistemas de Produçao, Electronic version, November, 2007.Available at,:http://sistemasde produçao.cnptia.embrapa.br/Fontes HTML/ Coco /Acultura do coqueiro/aspectos.htm. Accessed December, 1012

CHAN, E.; ELEVITCH, C. R. Cocosnucifera (coconut). In: ELEVITCH, C. R. (ed.). Species profiles for Pacific.**Islandagroforestry**. Permanent Agriculture Resources (PAR), Holualoa, 2006, 27p.

CHANG. C.L. WU. R.T. Quantification of (+) catechin and (-) epicatechin in coconut water by LC-MS. Food Chemistry, 126, 710-717.2011.

CHAVES, J.B.P. **Sensory analysis: history and development**. Viçosa: University Press, p.31, 1993.

CHITARRA, M. F. I.; CHITARRA, A. B. Post-harvest of fruit and vegetables: physiology and handling. Lavras: ESAL/FAEP, 2001. 289 p

CHITARRA, M. F. I.; CHITARRA, B. A. **Post-harvest of fruit and vegetables: physiology and handling.** 2. ed. Lavras: UFLA, 2005. 785 p.

CRIBB, AY. **Agro-industrial verticalization and cooperative management: a comparative analysis of agro-industrial business alternatives in the coconut chain of the Northern**

Fluminense Region. Rio de Janeiro: Embrapa Agroindustria de Alimentos. (Research Project) n.p.2006.

CRIBB, AY. 2008. **Evaluation of the economic, social and environmental impacts of the technology for preserving green coconut water by refrigeration and freezing** - Base Year 2007. Rio de Janeiro: Embrapa Agroindustria de Alimentos. 13p.

CRUZ, A. G.; CENCI, S. A.; MAIA, M. C. A. Prerequisites for implementing the HACCP system in a minimally processed lettuce line. **Ciência e Tecnologia de Alimentos,** Campinas, v. 26, n. 1, p. 104-109, jan./mar. 2006.

CDC. **Surveillance for Foodborne-Disease Outbreaks** - United States, 1998-2002.Center for Disease Control and Prevention Morbidity and Mortality Weekly Report (MMWR), 55, 134.2006.

DILLON, V.M. Yeasts and moulds associated with meat and meat products. In: The Microbiology of Meat and Poultry, ed. A. **Davies and R. Board, Blackie Academic and Professional,** London, p.85-117, 1998.

DUTCOSKY S. D. **Anàlise Sensorial de Alimentos,** Editora Champagnat 2^a ed, 2007 239 p.

DUREK, C. M. **Verification of good manufacturing practices in dairies registered with the Federal Inspection Service (SIF).** 97 f. Dissertation (Master's Degree in Veterinary Sciences) - Agricultural Sciences Sector, Federal University of Paranà, Curitiba, 2005.

EFSA (2010), "The Community Summary Report on Trends and Sources of Zoonoses, Zoonotic Agents and foodborne outbreaks in the European Union in 2008", The EFSA Journal, 1496, April (available at: http://www.efsa.europa.eu/en/scdocs/scdoc/130r.htm, retrieved 19.07.2010).

EGAN, M. B.; RAATS, M. M.; GRUBB, S. M.; EVES, A.; LUMBERS, M. L.; DEAN, M. S.; ADAMS, M. R. A review of food safety and food hygiene training studies in the commercial sector.**FoodControl**, v. 18, n. 10, p. 1180-1190, Oct. 2007

EMBRAPA. Refrigerated green coconut water 2005. Cabral, Lourdes Maria, Corrêa, L, Edmar das Mercês Penha, Virginia Martins da Matta. - Brasilia, DF: Embrapa **Informaçâo Tecnológica,** 2005.34 p.

EMBRAPA. **Evolution of coconut production in Brazil and international trade: 2010 panorama.** Carlos Roberto Martins, Luciano Alves de Jesus Jûnior - Aracaju: Embrapa Tabuleiros Costeiros, 2011.

EMBRAPA. Coconut will be the subject of a course and symposium in SE. Available at:<http://www.embrapa.gov.br/imprensa/noticias/Acesso on: 21 jul. 2010

FAGUNDES NETO, U. et al. Coconut water - variations in its composition during the ripening process. **Jornal de pediatria**, Rio de Janeiro, v.65, n.1/2, p. 17-21, jan./feb. 1989.

FAGUNDES NETO, U. et al. Negative findings for use of coconut water as an oral rehydration solution in childhood diarrhea. **Journal of the American College of Nutrition**, v.12, n.2, p.190-193, 1993.

FARIAS, J. M. de.; ALVES, R. E.; MACIEL, V. T. Cold damage in green Anao coconut fruits during refrigerated storage. In: ENCONTRO UNIVERSITARIO DE INICIAÇÂO À PESQUISA. 25, 2006, Fortaleza. **Proceedings...** Fortaleza: Universidade Federal do Cearà, 2006. 1 CD ROM.

FOOD AND AGRICULTURE ORGANIZATION OF THE UNITED NATIONS; WORLD HEALTH ORGANIZATION (FAO/WHO).FAO/WHO Guidelines for Governments on the application of the HACCP system in small and/or less developed food companies. Rome: FAO/WHO, 2007. p. 83.

FAO, Food and Agriculture Organization of the United Nations, <u>Available</u> atwww.fao.org.Dados from Faostat, 2007.

FAO. **World production.** 2009. Available: <http://faostat.fao.org/sote/567/defaut. aspx.ancor>.Accessed May 10, 2013.

FAO 2011. **World Production**. Available at: <www.faostat.org.br>. Accessed on: January 2012.

FOLHAON LINE. Electronic publication .
Available at
http://www1.folha.uol.com.br/folha/dinheiro/ult91u81792.shtml. Accessed on: April 3, 2004.

FONTES, H. R.; RIBEIRO, F. E.; FERNANDES, M. F. Coconut production: technical aspects. Brasilia, Embrapa: **Informaçao Tecnològica**, 2003. 105 p.

FERNANDES, I. L.; MOURA, L. S.; SILVA, R. C.; SANTOS, G. A.; SOUSA, A. R; COSTA, O. S.; OLIVEIRA, S. B. Comparison of the physical and chemical characteristics of natural and industrialized green coconut water. In: BRAZILIAN CONGRESS OF CHEMISTRY:

ENVIRONMENT AND ENERGY. 51, 2011, Sao Luiz. **Proceedings... Sao Luiz**: Federal University of Goiás, 2011. 1 CD ROM.

FERRAZ, M. S. Brazil is the world's third largest fruit producer. Online Magazine Brasil Alimentos. Sao Paulo, Aug. 2009.

FELLOWS, P. J. **Food Processing Technology.** Principles and Practice. 2nd Edition, Porto Alegre: Ed. Artmed, 2006.

F I. P. A. Newsletter **of the Federation of Portuguese Agri-Food Industries**. Food and lifestyles. V. 03, n° 52, p. 02, 2004.

FRANCO, B. D. G. M.; LANDGRAF. **Microbiologia dos Alimentos**, 2 ed. Sao Paulo: Editora Atheneu, 2007.

FRANCO, B. D. G. M.; LANDGRAF, M. **Microbiologia dos Alimentos**. Sao Paulo: Atheneu, 2003.182p

FRASSETI, J.; TÓRTORA, J. C. O.; GREGÓRIO, S. R. Acceptance of fresh and processed coconut water. In: **BRAZILIAN CONGRESS OF FOOD SCIENCE AND TECHNOLOGY**, 17. Fortaleza, 2000. Proceedings. Fortaleza: Brazilian Society of Food Science and Technology, 2000. v.1, p. 3.87.

GARBUTT, J. (1997). **Essentials of Food Microbiology.**London, UK: Arnold.

GARCIA, J. L. M. Raw material. In: MEDINA, J. C. et al. **Coconut: from cultivation to processing and marketing.** Sao Paulo: ITAL, 1980. chap. 2, p. 173-182.

GAVA, J. A. **Principios da tecnologia de alimentos.** Sao Paulo: Nobel. 1998. 284 p.

GAIVA, H. N. et al. Growth of an annual cultivar and four inter-varietal hybrids of coconut palm in the swampy region of Poconé, MT. In: BRAZILIAN CONGRESS

DEFRUTICULTURE, XVIII, 2004, Santa Catarina, **Anais... Santa Catarina**: CBF, 2004, 4 p.CD ROM.

GERMANO, P.M.L.; GERMANO, M.I.S. **Higiene e Vigilância Sanitària de Alimentos**. Sao Paulo: Varela Editora e Livraria Ltda. 629 p. 2001.

GERMANO, P.M.L; GERMANO, M.I.S.**Higiene e Vigilância Sanitaria de Alimentos.**Sao Paulo;Varela.655p.2003.

GERMANO, P. M. L.; GERMANO, M. I. S. **Higiene e Vigilância Sanitària de Alimentos.** 3. ed. Sao Paulo: Editora Manole, 2008. p. 13-20, 53-92, 317-323, 328-333.

GOMES, F. P. Cultivation of the Anao coconut palm: climatic and nutritional requirements. In. ZAMBOLIM, L. (Ed). **Integrated management: integrated production of tropical fruit, pests and diseases.** Viçosa: UFV, 2003. p. 95 - 111. (chap. 4).

GREIG, J. D.; LEE,M.B.(2009).Enteric outbreaks in long - term care facilities and recommendations for prevention: a review. **EpidemiologyandInfection**, 177, 145-155.

HOFFMANN, A. et al. Influence of temperature and polyethylene on the storage of serrana guava fruits (*Feijoasellomiona Berg.*) Revista **Scientia Agricola,** Lavras, v.5, n. 3, p. 563-568, 1994.

HOFFMAN, F. L.; COELHO, A. R., MANSOR, A. P., TAKAHASHI, C. M., VINTURIM, T. M. Microbiological quality of coconut water samples sold by street vendors in the city of Sao José do Rio Preto - SP. Revista Higiene Alimentar, Sao Paulo, v. 16, n.97, p.87-92, 2002.

IBGE. Brazilian Institute of Geography and Statistics.Municipal agricultural production. Rio de Janeiro, v. 34, p.1-69, 2007.

BRAZILIAN Institute of Geography and STATISTICS - IBGE. Available at: http://www.ibge.gov.br/cidadesat/topwindow.htm?1 . Accessed on: 27/01/2009.

IDESP. Institute for the Economic and Social Development of Parà. Parà State Government. Coconut culture in Parà. **Estudos Paranaenses**, Belém, v.45, jan. 1975. 86p.

INNOVATIVE. HACCP Food Safety System. **Internal Company Document.** 20pp. 2009.

JACKSON,V.; BLAIR, I.S.; MCDOWELL, D.A.; KENNEDY, J.; BOLTON, D.J. The incidence of significant foodborne pathogens in domestic refrigerators.**Food Control**, v. 18, p. 346-351, 2007

JAY, J.M. Indicators of Food Microbial Quality and Safety. In: **Modern Food Microbiology,** chapter 20, 6th ed., p.387-409, 2000

JAY, J.M., Loessner, M.J. Golden, D. A. **Modern Food Microbiology** (7^a ed.).United States of America, USA: Springer.(2005).

JAYALEKSHMY, A. et al. Changes in the chemical composition of coconut water during maturation. **Oléagineux**, v.43, n.11, p. 409-412, Nov. 1988.

JAYALEKSHMY, A.; ARUMAGHAN, C.; NARAYANAN, S.; MATHIEW, A. G. Changes in the chemical composition of coconut water during maturation.**Journal Food Science Technology,** v. 23, n. 4, p. 203-207, 1986.

KADER, A. A. et al. **Postharvest Technology of Horticultural Crops.**California: Universityof California, 1985. 148 p.

KADER A.A. Postharvest technology of horticultural crops.2.ed. Division of Agriculture and Natural Resources. Davis: University of California, n. 3311, 295p, 1992.

KLUGE, R.A.; NACHTIGAL, J. C., FACHINELLO, J. C.,BILHALVA, A. B. **Fisiologia e manejo pós-vestita de frutas de clima temperado**. 2. Ed. Campinas: Editora Rural, 2002. 214p.

KONEMAN, E. W. ; Allen, S. D. ; Janda, W. M. Schreckenberger, P. C.; Winn, w c. Microbiological Diagnosis: High Color Text. 5^a ed.; Ed. MEDSI, 2008.

KWIATKOWSKI, A.; OLIVEIRA, D. M.; CLEMENTE, E. Enzymatic activity and physicochemical parameters of coconut water harvested at different stages of development and climatic season. **Revista Brasileira de Fruticultura.** v. 34, n. 2, Jaboticabal, jun. 2012.

LEBER, A. S. M. L. **Evaluation of the stability of coconut water (*Cocos nucifera*) in polyethylene terephthalate (PET) bottles stored frozen and refrigerated.** 2001. 151f. Dissertation (Master's Degree in Food Technology) - State University of Campinas, Campinas, 2001.

LEBER, A. S.; FARIA J. A. F. Green coconut: characteristics and post-harvest care. **Revista frutas & Legumes**, n. 18, p. 36-38, Mar/Apr. 2003.

LIKAR, K. JEVSNIK, M. (2006).Coldchainmaintaining in food trade.**FoodControl**, 17, 108113.

LIRA, A. L. **Commercial sterilization process of green coconut water using ceramic membranes.** 2010. 141f. Dissertation (Master's in Process Engineering) - Federal University of Campina Grande, 2010.

LOIOLA, C. M. **Behavior of coconut palm cultivars (*Cocos nuciferaL.*) in different agroecological conditions of the coastal tabuleiros of northeastern Brazil.** 2009. 74p. Dissertation (Master's Degree in Agroecosystems) - Federal University of Sergipe, Sao Cristòvao.

LOVATTI, R. C. C. Food quality management: a practical approach. **Higiene Alimentar**, Sao Paulo, v. 18, n. 125, p. 90-93, 2004.

MACHADO, S.S. et al. Physical and physico-chemical characterization of yellow passion fruit from the region of Jaguaquara-Bahia. In: **XVIII SBCTA Congress**, August 4-7, 2008, Porto Alegre-RS. Proceedings. 1 CR-ROM.

MACIEL, V. T.; GOMES FILHO, E.; ALVES, R. E.; FARIAS, J. M. de.; MOURA, C. F. H.;SOUSA, H. U. de. Activities of antioxidative enzymes in the water of coconuts at different stages of development. In: CONGRESSO BRASILEIRO DE FRUTICULTURA, 20.,Vitoria, **Anais...** Vitoria: Sociedade Brasileira de Fruticultura, 2008. 4p. 1 CD ROM.

MACIEL, V.T. GOMESFILHO, E.; ALVES, R. E.; FARIAS ,J. M.; SOUZA, H. U. Physical characterization of the fruit of six Anao coconut cultivars at different stages of development. **Revista Brasileira de Ciências Agrarias**, Recife, v 4,n 4, p.395 - 398, 2009.

MAGALHÂES, M.P.et al. Preservation of green coconut water by membrane filtration.**Ciência e Tecnologia de Alimentos**, v.25, n. 1, p. 72-77,2005.

MARTINS, C.R.; TAVARES, J.C.; VASCONCELOS, G.C. de. Post-harvest of temperate climate fruits - Parameters for monitoring ripening - UFPEL, 2005.

MARQUES, A.L.V. Agua de coco - um produto tropical de mil e uma utilidades. **Revista Alternativa**, v.1, n.1, p 9-10, 1976.

MARQUES, L. M. P. C.; GALLI, V. Coconut water: nutritional and functional properties and commercialization. **V Semana de Tecnologia em Alimentos**, Ponta Grossa, v. 2, n. 1, p. 1-10, 2007.

MASSAGUER, P. R. Aseptic treatment of food: Focus on the nutritional and sensory quality of food, 2007. http//www.uncnet.br/arquivos/noticias/Apares Pilar. pdf. Accessed: February 2012

MINISTRY OF AGRICULTURE, LIVESTOCK AND SUPPLY - MAP. Brazilian agribusiness exports - total: ranking by 2010 values.

MEILGAARD, M.; CIVILLE, V.; CARR, B.T. Sensory Evaluation Techniques. Boca **Raton -FL**: CRC Press, Inc. 1987. 281 p.

MEDINA, J. C. Process: cultivation - varieties; products, characteristics and utilization - distilled beverages - coconut water - From cultivation to processing and commercialization. Sao Paulo: ITAL, 1980. 252 p. **Tropical Fruit Series**, n. 5.

NARAYAN, K. K.; DEO, J. V.; ABANI, M. C. Natural tritiumlevels in tender andripe coconut fruit (Cocosnucifera L.): a preliminary examination. **The Science of the Total Environment,** v. 256, p. 233-237, 2000.

NERY, M.V.S.; BEZERRA, V.S.; LOBATO, M.S.A. Physico-chemical evaluation of banana coconut grown in the state of Amapà. In: XVII Congresso Brasileiro de Fruticultura Brasileira, Nov. 2002, Belém-PA. **Proceedings...** 1 CD-ROM

RURAL NORTHEAST. Ceará stands out in exports. Accessed on: July 23, 2009.

NOGUEIRA, A. L. et al. Sensory evaluation of coconut water (*Cocos nucifera L.*) in natura and processed. **Revista de Biologia e Ciências da Terra**, Minas Gerais, v. 2, p. 10 - 14 2004.

NOGUEIRA, A. L. C. et al. Sensory evaluation of coconut water (Cocos nucifera L) in natura and processed. **Revista Biológica Ciências da Terra**, 2004; v.4, n.2, p. 1-5.

OLIVEIRA C. A; GERMANO P.M. Estudo da ocorrência de enteroparasitas em hortaliças comercializadas na regiao metropolitana de Sao Paulo, SP, Brasil- Pesquisa de helmintos. **Revista de Saùde Publica** 26.p. 283- 289 1992.

OLIVEIRA, M. E. B., BASTOS, M. S. R., FEITOSA, T. Avaliação de parâmetros de qualidade fisico-quimicos de pulpas congelados de acerola, cajà e caju. **Ciência e Tecnologia de Alimentos,** Campina Grande, v.19, n.3, p.326-332, 1999.

OLIVEIRA, F. M. N.; Figueiredo, R. M. F. de; Queiroz, A. J. de M. Comparative analysis of whole, formulated and powdered pitanga pulp. **Revista Brasileira de Produtos Agroindustriais**, v.8, p.25-33, 2006.

OLIVEIRA, R. C.; BARROS, S. T. D.; ROSSI, R. M. Application of Bayesian methodology to the rheological study of grape pulp. **Revista Brasileira de Produtos Agroindustriais**. Campina Grande, 2009. 8 p.

PAN-AMERICAN HEALTH ORGANIZATION (OPAS). HACCP: An essential tool for food safety. Buenos Aires: PAHO/INPAAZ, 2003. 401p.

PALANIAPPAN, S.; SASTRY, S.K. Electrical conductivity of selected solid foods during ohmicheating.**Journal of Food Process Engineering**, v.14, p. 2321-2360, 1991.

PAIS, E. Food Hygiene and Safety in a Pizzeria: Statistical Temperature Control. 2007.

PARANA. Paranâ State Health Department. Food outbreak. Available at http//www.saude.pr.gov.br/CSA/SURTO_alimentar/index.html.Acesso on: April 24, 2012

PENHA, E. M.; CABRAL, L. M. C.; MATTA, V. M. Agua de coco. In. VENTURNI FILHO, W. G. (Coord.). **Tecnologia de Bebidas: matéria-prima, processamento, BPF/APPCC e legislaçao.** Sao Paulo: Edgar Blucher, 2005. p.103-118. (ch. 5).

PETRUS, R. R.; FARIA J. A. F. Processing and stability evaluation of isotonic drinks in plastic bottles. Ciência e Tecnologia de Alimentos, vol 25, n°3, Campinas, Jul/set. 2005.

PINHEIRO, R.V.R.; MARTELETO, L.O.; SOUZA, A.C.G. de; CASALI, W.D.; CONDÉ, A.R. Productivity and fruit quality of ten varieties of guava, in Visconde do Rio Branco, Minas Gerais, aiming at natural consumption and industrialization. **Revista Ceres**, Viçosa, v.31, p.360-387, 1984.

PINHEIRO, A. M.; MACHADO, P. H.; COSTA, J. M. C.; MAIA, G. A.; FERNANDES, A. G.; RODRIGUES; M. C. P.; HERNANDEZ, F. F. H. Chemical, physico-chemical, microbiological and sensory characterization of different brands of coconut water obtained by the aseptic process. **Revista Ciência Agronòmica**, Fortaleza, v. 36, n° 2, p. 209 - 214 .2005.

POWELL, A. W. G. Shrink-wrap of avocados in combination with waxing and fungicide. South Africa Avocado Grower's Association Yearbook, n. 11, p. 39-40, 1988.

RANIERI, M. Agua de coco: um mercado em crescimento. Revista Engarrafador Moderno, Sao Paulo ,n 74,P.24-28,April 2000.

REGULATION (EC) No. ^{0} 852/2004 of the European Parliament and of the Council of April 29, 2004. lays down general rules for food business operators concerning the hygiene of foodstuffs.

Official Journal of the European Union: L 139, 30.04.2004.

RESENDE, J. M, de QUEIROZ M. R, SOARES A. G, BOTREL N, CABRAL. L. M. C, Jùnior JS. Physical, physico-chemical and chemical characterization of "green dwarf" coconut water coated with different biofilms. In: Congresso Brasileiro de Fruticultura, 20, 2008, **Anais...**Vitória, ES. Vitória, ES: Sociedade Brasileira de Fruticultura, 2008.

RESENDE, J. M.; VILAS BOAS, E. V. B.; CHITARRA, M. I. F. Use of modified atmosphere in the post-harvest preservation of yellow passion fruit. **Revista Ciência e Agrotecnologia,** Lavras, v. 25, n. 1, p. 159-168, 2001.

RESENDE, J. M. **Biodegradable coatings for the preservation of 'Anâo Verde' coconut.** 2007, 200p. Thesis (Doctorate in Agricultural Engineering). State University of Campinas. Faculty of Agricultural Engineering, Campinas.

RODRIGUES, M.I.; LEMMA, A. F.; Planning experiments and optimizing processes: a sequential planning strategy, 1^a. Ed. Campinas, SP: Casa do Pao. Editora, 2005.326p.

RODRIGUES, C.S. **Microbiological contamination in lettuce and cabbage sold in retail stores in Brasilia-** DF. 29p. Graduation Monograph. July 2007.

ROSA. M. F; ABREU, F. A. P, **Agua-de-coco Métodos de conservaçâo.** Document No. 37. Fortaleza, CE: Embrapa Agroindùstria Tropical. Fortaleza, 2000.

ROSA, M. F.; ABREU, F. A. P. Conventional processes for preserving coconut water. In: ARAGÂO, W. M. **Coconut:** post-harvest. Brasilia, DF: Embrapa Informaçâo Tecnològica, 2002. P. 52-53.

SALGUEIRO, C. C. M. **The applications of coconut water. Infococo.** Fortaleza: Grupo de Coco do Cearà, 2001.

SANTOS. J. E. F.; TEXEIRA. L. E. B.; MOREIRA. I. S.; SOUZA. F. C.; CASTRO. D. S. Microbiological evaluation of coconut water sold by street vendors in Juazeiro do Norte-CE, Revista Verde, v.8, n. 2, p. 23-26, April-June 2013.

SANTANA, L. R.R. et al. Sensory profile of peach-flavored light yogurt. **Revista Ciência e Tecnologia de Alimentos,** v. 22, n.1, p 23-36, 2006.

SAAT, M.; SINGH, R.; SIRISINGHE, R. G.; NAWAWI, M.Rehydration after exercise with fresh young coconut water,carbohydrate-electrolyte beverage and plain water. **JournalofPhysiologicalAnthropology,** Japan, v.21, n.2, p.93-104, 2002.

SANTOS, G. A.; BATUGAL, P. A.; OTHAM, A.; BAUDOWIN, L. E.; LABOUISSE, J. P.

Manual on standardized Research Techniques in coconut breeding. IPGRI, 1996, 45p.

SANTOS, S. A. et al. Microbiological and physico-chemical analysis of coconut water sold in the city of Aracajù - SE, IV **Latin American Symposium on Food Science**, UNICAMP - SP, 2001.

SANTOS dos, E. C. **Refrigerated storage under modified atmosphere of fresh and minimally processed Anao Verde coconut fruits**. Dissertation (Master's Degree in Plant Science). Federal University of Paraiba (UFPB), Areia. 2003. 73 f

SANTOS FILHA, M. E. C. **Quality and post-harvest preservation of fruit from six cultivars of Anao coconut palm.** Dissertation (Master's Degree in Plant Science) - Universidade Federal Rural do Semiàrido, Mossoró. 124f 2006.

SANTOS, J.; RIBEIRO, G. A. Microbiological evaluation of fresh orange juices sold in the city of Pelotas, RS. **Revista Higiene Alimentar**, Sao Paulo, v.20, n.147,p.40-44,2006.

SANTOSO, U. et al. Nutrientcompositionofkopyorcoconuts (Cocos nucifera L.). **Food Chemistry**, v.57, n.2, p. 299-304, 1996.

SEREJO, T.T. NEVES, M.A. BRITO N.M; **Microbiological Quality of Coconut Water Sold by Street Vendors in the City of Sao Luis - MA**. Federal Institute of Education, Science and Technology of Maranhao - IFMA, 2010.

SHEID, G.A. **Sensory and physico-chemical evaluation of Italian salami with different concentrations of cloves (Eugenia caryophyllus).** 2001. 94 f. Dissertation (Master's Degree in Food Science and Technology) - Federal University of Viçosa, 2001.

SILVA, A. S.; ARAUJO, M. V.; ALVES, R. E. Harvest point of green coconuts for fresh consumption. I Encontro de iniciaçao Cientifica da Embrapa Agroindùstria tropical, Fortaleza- CE, 2003.

SILVA, F. A. S.; AZEVEDO, C. A. V. Version of the Assistat computer program for the Windows operating system. **Revista Brasileira de Produtos Agroindustriais.** Campina Grande, v. 4, n. 1, p. 71-78, 2009.

SILVA, G.A.S. V.J.N.; L.F.F. **Physical and chemical characterization of fresh and industrialized coconut water sold in the Paraíba sertao.** Federal University of Campina Grande-UFCG-Pombal/PB. 2010

SILVA, D. L. V.; ALVES, R. E. ; FIGUEIREDO, R. W.; MACIEL, V. T. et al. 2009.Physical, physico-chemical and sensory characteristics of water from green dwarf coconut palm fruits from conventional and organic production.**Ciencia Agrotecnica**, 33 (4): 10791084.

SILVA JÙNIOR, E. A. **Manual de controle higiênico-sanitârio em serviços de alimentaçao.** 6

ed. ; Sao Paulo: Varela, 2005. 29 p.

SILVA, J. L. A.; DANTAS, F. A. V.; SILVA, F. C. Microbiological quality of coconut water sold in the municipality of Currais Novos, RN. **Revista Holos**, n. 25, v. 3, p. 3435, 2009.

SIVASANKAR, B. (2004). Food Processing and Preservation. India: PHI Learning Pvt. Ltd.

SOARES, A. A. L. J. **Physico-Chemical and Bromatological Evaluation of the Pulp of *Spondia spurpurea* L (ciriguela) in the Central Semi-Arid Region of Paraiba.** PATOS-PB-BRAZIL, FEBRUARY - 2011. Dissertation: Postgraduate Program in Forestry Sciences, Federal University of Campina Grande.

SOUZA, J. M. **Microbiological quality of semi-ready pizza doughs: critical points in production and marketing.** (Master's Degree in Food Science). Federal University of Minas Gerais, Belo Horizonte, 1997.

SOUZA de, V. A. B. et al. Evaluation of Anao coconut cultivars in the Baixo Parnaiba Piauiense micro-region: vegetative development characteristics. In: CONGRESSO BRASILEIRO DE FRUTICULTURA, XVII, 2002, Belém, **Anais...** Parà: CBF, 2002, 5 p.CD ROM.

SOUZA, S.O. et al. Physico-chemical changes in coconut water during fruit development. In: XVII Congresso Brasileiro de Fruticultura Brasileira, nov. 2002, Belém-PA. Proceedings...1 CD-ROM.

SOUZA, C. D. D. **Thermal Regeneration of Commercial Clays for Reuse in Soybean Oil Clarification.** Federal University of Santa Catarina Postgraduate Course in Chemical Engineering Integrated Technologies Laboratory. Department of Chemical Engineering and Food Engineering Florianópolis-SC, February 2002, 99p.

SOUSA, J. P. de.; PRAÇA, E. F.; ALVES, R. E.; NETO, F. B.; DANTAS, F. F.; Influence of refrigerated storage in association with modified atmosphere by plastic films on the quality of 'Tommy Atkins' mangoes. **Revista Brasileira de Fruticultura**, Jaboticabal, v. 24, n. 3, p. 665-668, 2002.

SOUZA, E.L.; SILVA, C.A; SOUSA, C.P. Sanitary quality of equipment, surfaces, water and hands of handlers of some establishments that sell food in the city of Joao Pessoa, PB. **Revista Higiene Alimentar**, v.18, n. 116, p. 98-102, 2004.

SMITH, J.P., Daifas, D. P., El-Khoury, W., Koukoutsis, J., El-Khoury, A. (2004). Shelf Life and Safety Concerns of Bakery Products-A Review.**Critical Reviews in Food Science and Nutrition,** 44, 19-55.

SCHMIDT, F. L. et al. 2004. Good manufacturing practices and application of the hazard analysis and critical control points system in coconut water processing. **Food Hygiene Magazine**, v. 18, n. 121, p.65-76.

SHAMES, I. H. **Fluid Mechanics** - volume 1. Sao Paulo: Editora Edgard Blücher,. ISBN:852-120-170-2.1999. 192 p

SHIMIZU, M. K. et al. Characterization of the harvest point of anao-green coconut palm fruits (*Cocos nuciferaL.*) in the Baixada de Sepetiba /RJ Region - Preliminary Results. In: CONGRESSO BRASILEIRO DE FRUTICULTURA, 15., 2002, Belém. **Proceedings...** Belém: SBF, 2002. 4 p. 1 CD.

SREBERNICH, S.M. **Physical and chemical characterization of coconut fruit water (*Cocos nucifera*), giant varieties and hybrid PB-121, with a view to developing a drink with characteristics close to those of coconut water.** 1998. 189 f. Thesis (Doctorate in Food Technology) - Faculty of Food Engineering, State University of Campinas, Sao Paulo, 1998.

SREBERNICH, S.M.; MORETTI, R.H.; CARVALHO, C.R.L. Determination of sugars in coconut water from the hybrid cultivar PB 121 (West African giant x Malaysian yellow dwarf). In: XVII Brazilian Congress of Food Science and Technology, Aug. 2000, Fortaleza-Ceará. **Proceedings...** Book of abstracts, v.2, p. 5.72.

STUMBO, C. R. **Thermobacteriology in Food Processing.** Academic Press Inc. New York, N. Y, 1973. 263p.

TAN, T.C., CHENG, L.H., BHAT, R., RUSUL, G., EASA, A. M. Composition, physicochemical properties and thermal inactivation kinetics of plyphenyl oxidase and peroxidase from coconut (Cocos nucifera) water obtained from inmature, mature and overly- mature coconut, **Food Chemistry,** Penang Malaysia, n.142, p.121-180, Jul, 2013.

TERUEL, BARBARA J.M. Cooling technologies for fruit and vegetables. **Revista. Brasileira de Agrociência,** Pelotas, v.14, n.2, p.199-220, apr- jun, 2008.

TSAY, L. M. Effects of storage temperature on the quality of sugar apple. **Journal Japan Society of Cold Preservation of Food**, Tokyo, v.14, p.6-7, 1988.

ALL FRUIT. Information on coconut production. Available at <http://www.todafruta.com.br/portal/icNoticiaAberta. Accessed: Feb. 10, 2011.

TAVARES, M. et al. Study of the chemical composition of -Anao Verde coconut water at different stages of maturation. In: BRAZILIAN CONGRESS OF FOOD SCIENCE AND TECHNOLOGY, 16, 1998. Rio de Janeiro. **Proceedings...** Rio de Janeiro: SBCTA, 1998. CD-ROM.

VASCONCELOS, A. R. D. **Use of calcium chloride and modified atmosphere in the**

preservation of persimmon cv. Fuyu. Dissertation (Master's Degree in Food Sciences), Federal University of Lavras (UFLA). Lavras, 2000, 85 p.

VIANA, F.M.P., UCHÔA, C.N., VIEIRA, I.G.P., FREIRE, F. C. O., SARAIVA, H.A.O., MENDES, F.N.P. Minimal processing, modified atmosphere, chemical products and cooling to control post-harvest basal rot of fresh green coconut fruits (*Cocosnucifera*).**Summa Phytopathologica**, v.34,n.4. p. 326-331, 2008.

WALTER, E. H. M.; KABUKI, D. Y.; ESPER, L. M. R.; Sant' Ana, A. S .; KUAYE, A. Y. Modelling the growth of Listeria monocytogenes in fresh green coconut (Cococs nucifera L.) water. **Food Microbiology**, v. 26, p. 653- 657, 2009.

WANG, C. Y. Effect of Preharvest factors affecting on postharvest quality: Introduction of the Colloquium. **Hort Science**, Betts ville, v. 32, n. 5. p. 807 - 811, aug. 1997.

VERLINDER, B. E.; NICOLAI, B.M. Fresh-cut fruits and vegetables.**ActaHorticulturae**, v.5, n. 18, 2000.

WESTON, L. A.; BARTH, M. M. Preharvest factors affecting postharvest quality vegetables.**HortScience.**V.32, n. 5, p. 812-815, 1997.

WHO. Drug-resistant Salmonella. Fact sheet n°139. Revised April 2005. Available at: http://www.who.int/mediacentre/factsheets/fs139/en/

8. APPENDIX A

Sensory sheet

Name:___________________________________

Date:_______

Please taste the sample by giving a score from 1 to 9 for each characteristic of the industrialized coconut water you evaluated, based on how much you liked or disliked the product, using the scale below:

9. I liked it a lot
8. I liked it a lot
7. I liked it moderately
6. I liked it slightly
5. I didn't like it or dislike it
4. Slightly dislike
3. Moderately disliked
2. I really dislike it
1. Strongly dislike

A)How do you rate the **APPEARANCE of** the product?

Sample	Scale value	Remarks

B)How do you rate the **AROMA of** the product?

Sample	Scale value	Remarks

C)How do you rate the **TASTE of** the product?

Aimostra	Scale value	Remarks

D) Mark the answer that best corresponds to your judgment **(attitude)**

5. I would certainly buy
4. I would probably buy
3. Maybe I would buy/maybe I wouldn't buy
2. I probably wouldn't buy it
1. I certainly wouldn't buy it

Aimostra	Scale value	Remarks

E) **Order** the samples according to your preference?
                FirstSecondThirdBedroom
Thank you for your cooperation!!!

Printed by Books on Demand GmbH, Norderstedt / Germany